3

FOREWORD

by Dr Patrick Moore

Co-founder of Greenpeace, Dr Patrick Moore is Chairman and Chief Scientist of

Greenspirit Strategies Ltd in Vancouver, Canada. Website: http://www.greenspiritstrategies.com

Today our foremost energy challenge is to meet increasing needs without adding to our environmental problems, notably global warming and air pollution.

Though there is wide and increasing consensus on the need to severely limit greenhouse gas emissions, a significant reduction seems unlikely, given our continued heavy reliance on fossil fuel consumption. Even UK environmentalist James Lovelock, who posited the Gaia theory that the Earth operates as a giant, self-regulating superorganism, now sees nuclear energy as key to our planet's future health. Lovelock says the first world behaves like an addicted smoker, distracted by short-term benefits and ignorant of long-term risk. "Civilization is in imminent danger," he warns, "and has to use nuclear – the one safe, available energy source – or suffer the pain soon to be inflicted by our outraged planet."

Yet environmental activists, notably Greenpeace and Friends of the Earth, continue to lobby against clean nuclear energy, and in favour of the band-aid Kyoto Treaty plus a string of unrealistic suggestions. We can agree that renewable energies, such as wind, geothermal and hydro are part of the solution. But nuclear energy is the only non-greenhouse gas-emitting power source that can effectively replace fossil fuels and satisfy global demand. The blind and anti-scientific opposition to this proposition goes back to the mid 1980s when Greenpeace and much of the environmental movement made a sharp turn to the political left and began adopting extreme agendas that abandoned science and logic in favour of emotion and sensationalism.

In the last two decades I have pursued the concept of sustainable development and sought to develop an environmental policy platform based on science, logic, and the recognition that more than six billion people need to survive and prosper, every day of the year. Environmental policies that ignore science can actually result in increased risk to human health and ecology. The zero-tolerance policy against nuclear energy that has been adopted by so many activist groups is a perfect example of this outcome. By scaring people into fearing atomic energy, they virtually lock us in to a future of increasing fossil fuel consumption.

That is why I am pleased to commend this book, effectively an eighth edition of a comprehensive introduction to nuclear power, with a scientific basis and pitch. That is where I believe discussion and public debate on the question – and energy policies generally – needs to begin and remain based.

Nuclear energy can play a number of significant roles in improving the quality of our environment while at the same time providing abundant energy for a growing population. First, as mentioned above, it can replace coal and natural gas for electricity production. Coal-fired power plants in the US alone produce nearly 10% of global CO_2 emissions. Under present scenarios, even with aggressive

The World Nuclear University Primer

Nuclear Energy

in the 21st Century

Ian Hore-Lacy

Previously published as **Nuclear Electricity**: 1st - 7th editions, 1978 - 2003

World Nuclear University Press

OCM 716/7807

Carlton House   22a St James's Square

London SW1Y 4JH   United Kingdom

Elsevier Inc.

30 Corporate Drive

Burlington, Massachusetts 01803   USA

World
Nuclear
University
Press

CONTENTS

growth in renewable technologies, coal and natural gas consumption will continue to increase rather than decrease. The only available technology that can reverse this trend is nuclear energy.

France, for example, now obtains over 75% of its electricity from nuclear plants. The other 20% is mostly hydroelectric, therefore making France's electrical production virtually greenhouse gas-free and pollution-free. If other countries had followed France's path, there wouldn't be as much of a climate change issue around power production as there is today.

Second, nuclear energy can be used to produce hydrogen for a future fuel cell-based transportation system. A nuclear plant can produce sufficient heat to split water into hydrogen and oxygen thermally. This is much more efficient than using electricity to split water. There are a lot of technical hurdles and the hydrogen economy may still be years away, but there is no other alternative to using fossil fuels for transportation in the offing. A conversion to hydrogen would not only solve greenhouse gas and pollution concerns, it would have considerable geo-political implications regarding energy security.

Third, nuclear energy can be used to desalinate seawater to provide water for drinking, industry and irrigation. A growing population, shrinking aquifers and increasing irrigation demand all add up to the need to make our own fresh water in the future. Nuclear can provide the energy to do it without causing pollution or greenhouse gas emissions.

Fourth, we will continue to use fossil fuels, hopefully at reduced levels, far into the future. As conventional supplies of oil diminish we will turn to the vast shale oil and oil sand deposits. This is already a growing industry in northern Canada where the oil sands contain as much proven supply as Saudi Arabia. But the oil costs more because it must be separated from the sand. This is done by burning large volumes of natural gas to make steam, then basically steam-cleaning the sand to get the oil. By using one fossil fuel to obtain another there are even more greenhouse gas emissions than from burning conventional oil supplies. One solution to this would be to use nuclear energy to make the steam, and electricity, to run these oil sand and shale oil projects. This would substantially reduce greenhouse gas emissions and air pollution.

There are about 440 nuclear power reactors operating in 30 countries, producing 16% of the world's electricity. This could be doubled or tripled if the political will were brought to bear on the issue of reducing fossil fuel consumption. I believe that the environment would benefit from moving in this direction. Let's hope the future takes us there.

ABOUT THE AUTHOR

Ian Hore-Lacy, MSc FACE, a former biology teacher, became General Manager of the Uranium Information Centre, Melbourne, in 1995 and Head of Communications for the World Nuclear Association, based in London, in 2001.

He joined the mining industry as an environmental scientist in 1974, where his responsibilities covered uranium production. From 1988-1993 he was Manager, Education and Environment, with CRA Limited (now Rio Tinto). Since 1976, he has written and published several books on environmental and mining topics, including *Responsible Dominion – a Christian Approach to Sustainable Development* (Regent College Publishing 2006). He is a contributor to Elsevier's *Dictionary of Energy* and *Encyclopedia of Energy* (both 2005). His particular interests range from the technical to the ethical and theological aspects of mineral resources, including their use in applications such as nuclear power. He has four adult children.

Early editions of this book owed their substance to Ron Hubery as co-author. Ron is a chemical engineer, now retired, who spent eight years working with the Australian Atomic Energy Commission (now the Australian Nuclear Science and Technology Organisation) on nuclear fuel cycles and reprocessing. He also worked at the uranium production centres of Rum Jungle and Mary Kathleen in Australia.

ACKNOWLEDGMENTS

This text builds on seven editions of *Nuclear Electricity*, 1978-2003, published (since 4th edition) by the Uranium Information Centre (UIC) in Melbourne, assisted at various stages by Atomic Energy of Canada Ltd and the World Nuclear Association. Without that basis and my colleagues at WNA since 2001, this book would have been a formidable and possibly futile undertaking.

Section 3.7 on Physics draws heavily on material written for UIC by Dr Alan Marks. Chapter 9 draws heavily on the introductory section of *Atomic Rise and Fall, the Australian Atomic Energy Commission 1953-1987*, by Clarence Hardy, Glen Haven, 1999; and the Russian part of it was largely contributed by Judith Perera. In all cases material is used with permission.

The front cover image was supplied by General Electric Co. and shows an Advanced Boiling Water Reactor (ABWR).

This book was designed by Brigita Praznik, Graphic Information Designer at the World Nuclear Association. Figures in this edition were by Sara Pavan.

INTRODUCTION

The context

There is a rapidly-increasing world demand for energy, and especially for electricity. Much of the electricity demand is for continuous, reliable supply on a large scale, which generally only fossil fuels and nuclear power can meet.

The fuel for nuclear power to make electricity is uranium, and uranium's only substantial non-weapons use is to power nuclear reactors. There are some 900 nuclear reactors operating today around the world. These include:

- about 260 small reactors, used for research and for producing isotopes for medicine and industry in 56 countries,
- over 220 small reactors powering about 150 ships, mostly submarines,
- some 440 larger reactors generating electricity in 30 countries.

Practically all of the uranium produced today goes into electricity production with a significant small proportion used for producing radioisotopes. In particular, uranium is generally used for base-load electricity. Here it competes with coal, and in recent years, natural gas.

Over the last 50 years nuclear energy has become a major source of the world's electricity. It now provides 16% of the world's total. It has the potential to contribute much more, especially if greenhouse concerns lead to a change in the relative economic advantage of nuclear electricity, emphasizing its ethical desirability. On top of that there is an emerging prospect of the Hydrogen Economy, with much transport eventually running on hydrogen. Just as nuclear power now produces electricity as an energy carrier, it is likely to produce much of the hydrogen, another energy carrier.

The uranium and nuclear power debate today is about options for producing electricity. None of those options are without some risk or side effects.

Since the first edition of this book's predecessor in 1978 – *Nuclear Electricity* – many of the expectations surrounding alternative energy sources have been shown to be unrealistic (as indeed have some of those for nuclear energy). However, it is important that this return to reality does not lead to their neglect; such alternatives should continue to be developed, and applied where they are appropriate. In particular a great deal can be achieved by matching the location, scale and thermodynamic character of energy sources to particular energy needs. Such action should be a higher priority than merely expanding capacity to supply high-grade electrical energy where for example only low-grade heat is required, or using versatile gas to generate electricity on a large scale simply because the plant is cheaply and quickly built.

But when the question of utilizing nuclear energy arises, there are those who wish somehow to put the genie back in the bottle and to return to some pre-nuclear innocence. The debate in Europe is instructive: France gets over 75% of its electricity from nuclear power. It is the world's largest electricity exporter, and gains some EUR 2.5 billion per year from those exports. Next door is Italy, a major industrial country without any operating nuclear power plants. It is the world's largest net importer of electricity, and most of that comes ultimately from France. Elsewhere Germany and Sweden have nuclear phase out policies which are patently unrealistic.

The present and future roles of nuclear power are not limited to electricity, and hence the expanded scope of this book beyond *Nuclear Electricity*. The large potential for nuclear heat to make hydrogen to fuel motor vehicles is perhaps the chief interest today, but nuclear energy for desalination (as potable water becomes a more valuable commodity), marine propulsion, space exploration, and research reactors to make radioisotopes are all encompassed in this book.

I anticipate that my grandchildren's generation will come to look upon weapons as simply an initial aberration of the nuclear age, rather than a major characteristic of it. Arguably the same is true of the bronze and iron ages, where weapons provided incentive for technological development which then became applied very widely.

Certainly, as Figure 1 graphically shows, we cannot indefinitely depend on fossil fuels as fully as we do today.

Figure 1: Consumption of fossil fuels

Source: Charles McCombie, NAGRA Bulletin # 29, 1997, based on Korff, 1992 (and probably M.K. Hubbert, 1969, who had the peak around 2100).

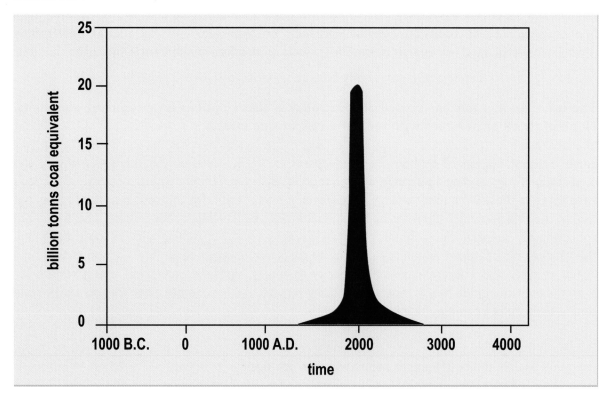

The book

Considerable effort has been made to include as much up-to-date and pertinent information as possible on generating electricity from nuclear energy, and on other uses of nuclear power. The figures quoted are conservative, and generalizations are intended to withstand rigorous scrutiny. The reader will not see many of the frequently repeated assertions from supporters or opponents of nuclear energy. The book does not enter into debate on social issues.

Since the first edition of *Nuclear Electricity*, the intention has been to get behind the controversies and selective arguments and present facts about energy demand and how it is met, in part, by nuclear power. Every form of energy production and conversion effects the environment and carries risks. Nuclear energy has its challenges, but these are frequently misunderstood and often misrepresented. Nuclear energy remains a safe, reliable, clean, and generally economic source of electricity. But many people do not see it that way.

This edition comes out at a time when the contrast between environmental concern focused on tangible indicators of pollution and global warming is beginning to stand in stark contrast to Romantic environmentalism, which is driven by mistrust of science and technology, and which demonizes nuclear power. Increasing evidence of the contribution to global warming from burning fossil fuels is countered by fearmongering often based on the 1986 Chernobyl disaster.

The introduction to the first edition of this book in the 1970s expressed the opinion that if more effort were put into improving the safety and effectiveness of commercial nuclear power, and correspondingly less into ideological battles with those who wished it had never been invented, then the world would be much better off. With Chernobyl nearly two decades behind us and the great improvements to safety in those plants which most needed it, plus the welcome recycling of military uranium into making electricity, it seems that we are now closer to that state of affairs.

Further information

Throughout the text, there are references to World Nuclear Association (WNA) Information Papers, which cover some of the issues found in the book in more detail. These papers can be read by visiting the WNA website at: http://www.world-nuclear.org. This site also offers up-to-date news, articles and reports on nuclear energy issues, as well as links to other sites with reliable information.

ENERGY USE

1.1 SOURCES OF ENERGY

All energy is derived ultimately from the elemental matter which comprises both the Sun and the Earth, formed in supernovae over 6 billion years ago. From the Sun we have both fossil fuels and the main contemporary renewable sources. From the elemental substance of the Earth we have uranium and geological heat.

The Sun warms our planet, and provides the light required for plants to grow. In past geological ages the Sun provided the same kind of energy inputs. Its energy was incorporated into the particular plant and animal life (biomass) from which were derived today's coal, oil and natural gas deposits – the all-important fossil fuels on which our civilization depends.

The only other ultimate energy source in the Earth is from the atoms of particular elements formed before the solar system itself. These are found today in the Earth's crust[1] and mantle.

The amount of energy per unit mass of an atom is dependent on the size of the atom: the minimum amount of energy per unit mass is contained within the medium-sized atoms (such as carbon and oxygen), whereas the greatest amount is contained in small atoms (such as hydrogen) or large atoms (such as uranium). Energy can therefore be released by combining small atoms to produce larger ones (fusion) or by splitting large atoms to produce medium sized atoms (fission). The tapping of this energy by nuclear fission or by nuclear fusion is one of the most important and contentious human achievements in history.

1.2 SUSTAINABILITY OF ENERGY

Much has been written since the early 1970s about the impending "world energy crisis", which was initially perceived as a crisis due to limited oil supplies. Today it is more a geopolitical crisis due to the location of supplies of oil and gas resources relative to demand for them. But finite supplies are still a factor, and Figure 1 (see Introduction) suggests the vital importance of conserving fossil fuel resources for future generations and the importance of sustainability.

Since the early 1970s the pressure has been to conserve crude oil supplies, but in the future it will increasingly be to reduce burning of all fossil fuels. Today global warming concerns drive this trend strongly. It is likely that coal will take over some of the roles of oil today, especially as a chemical feedstock. Sustainability of energy relates both to adequacy of supplies and the environmental effects of its use.

The importance of energy conservation is obvious, even in areas where so far fuels have been relatively cheap, and the need to limit carbon emissions lends emphasis to this. The levelling-out of overall energy demand in developed countries in recent decades is a result of increased energy efficiency. However, in developing countries growth in energy demand from a low starting point continually increases the pressures on resources worldwide, despite conservation initiatives (see Table 1).

Many people in developing nations aspire to the standard of living, mobility, agricultural productivity and industrialization characteristic of the developed countries. Fulfilling these hopes depends on the availability of abundant energy. Growth of the world's population from the present level of 6 billion people to a projected 8 billion in 2025, mostly in today's developing nations, increases the challenge.

---

[1] Uranium appears to have been formed in supernovae some 6.5 billion years ago, and though not common in the solar system has been concentrated in the Earth's crust at an average of c. 1.4 ppm. Heat from the radioactive decay of this uranium today drives the convection processes in the Earth's mantle and is vital to life.

Table 1: Growing electricity production – terawatt hours (TWh, or billion kWh)

	1990	2003	increase
OECD	7603	9938	31%
Non-OECD	4270	6307	148%
World	11873	16742	41%
Non-OECD:			
Former USSR	1727	1349	minus 22%
Africa	323	507	57%
Latin America	491	829	69%
Asia (exc China)	647	1433	121%
China	650	1943	199%
Middle East	236	553	134%

Source: OECD[2]/International Energy Agency (IEA) 2005, Energy Statistics of Non-OECD Countries, 2002-2003.
See also Figures 2 & 5.

1.3 ENERGY DEMAND

In industrialized countries energy demand derives from three major sectors:

- Domestic and commercial
- Industry and agriculture
- Transport

In many countries these each account for about one third of the energy demand, although the size of domestic demand depends very much on climate. In Australia, for example, domestic demand is relatively small, whereas in Canada it is extremely large because of the cold climate.

More specifically it is possible to identify demand for particular purposes within these sectors, such as the following:

- Low temperature heat (up to 110°C) for water and space heating in homes and industry
- High temperature heat (over 110°C) for industrial processes
- Lighting
- Motive power for factories, appliances and some public transport
- Mobile transport for public and private use

For some of these purposes there is a significant demand for energy in the form of electricity. Worldwide, electricity demand is increasing very rapidly, as illustrated in Table 1 and Figure 3. This is discussed further in section 2.1.

1.4 ENERGY SUPPLY

On the supply side, there are a number of primary energy sources available (see Figure 2). Derived from these primary sources are several secondary energy sources or carriers. These include, for example:

- Electricity – can be generated from many primary sources
- Hydrogen – mainly from natural gas or electrolysis of water
- Alcohols – from wood and other plant material
- Oil and gas – manufactured from coal

At this stage only electricity is of major importance as a secondary source, but hydrogen is expected to become significant in the future as a replacement for oil products. (see Chapter 6).

---

[2] Organization for Economic Cooperation and Development

Much energy demand can be met by more than one kind of energy supply. For instance, low temperature heat can be produced from any of the fossil fuels directly, from electricity, or from the Sun's radiant energy. Other demands such as mobile transport need to be supplied by portable fuels such as those derived from oil or gas. In the future, hydrogen is expected to become important in this role.

Both economic practicality and ethical considerations mean that versatile, easily portable energy sources such as oil and its derivatives are not usually squandered where other, more abundant fuels can be substituted.

Primary energy resources in different countries vary enormously. There are great differences in natural endowment, and this makes clear the importance of trade in energy, as indicated in Table 2. Different energy sources also yield different amounts of energy per unit mass or volume, as shown in Table 3 at the end of this Chapter.

1.5 CHANGES IN ENERGY DEMAND AND SUPPLY

The uneven world distribution of energy resources means that as energy consumption rises, international trade in energy must increase. Energy-poor countries find themselves dependent on supplies from energy-rich countries, as Table 2 illustrates. Because of the fundamental importance of energy in the industrial economy, importing countries are vulnerable politically and economically. Energy trade between regions is projected to double by 2030, and most will continue to be in the form of oil.

The best illustration of this vulnerability is the changing position of oil. Until the early 1970s, many countries had come to depend on oil because of its relatively low cost, and world oil production tripled between 1960 and 1973. But this suddenly changed as prices rose four-fold, and this was then followed by a further "oil crisis" in 1979. As a result, world oil consumption in 1986 was the same as that in 1973, despite a substantial rise in total primary energy consumption. Forecasts in 1972 had generally predicted a doubling of oil use in ten years.

Table 2: Energy use, with net imports and exports (petajoules) in 2002

	Own energy use – total primary energy supply	Net import	Net export
Australia	4730		5820
Canada	10470		5720
France	11140	5690	
Germany	14490	8840	
Japan	21650	17800	
Russia	25880		17184
Saudi Arabia	5020		14030
UK	9500		1260
USA	95880	26380	

Sources: *OECD/IEA 2004, Energy Balances of Non-OECD Countries, 2001-2002.*
OECD/IEA 2004, Energy Balances of OECD Countries, 2001-2002.

Figure 2: World primary energy demand

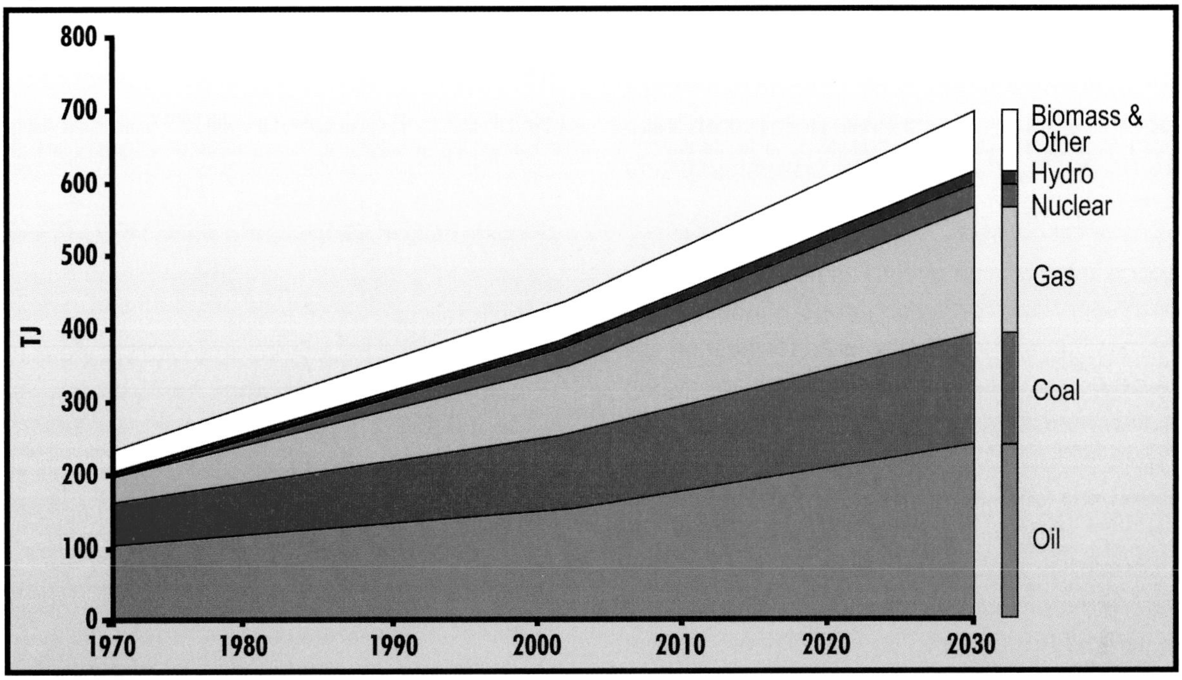

Source: OECD/IEA World Energy Outlook 2004.

Japan, for example, has few indigenous energy resources and little untapped hydroelectric potential. It suddenly found that escalating oil imports to supply three quarters of its total energy needs were not sustainable. Even the USA, originally self-sufficient in oil, found it difficult to pay for enough imported oil to offset declining domestic production, and today security of supply is a major factor in its foreign policy.

Problems of oil prices and supply in the 1970s brought about rapid changes in the production and use of other primary energy resources:

- Coal production and international trade in coal increased to substitute for some oil use. It is currently growing strongly again.

- Nuclear power for electricity generation was adopted or examined more closely by energy-deficient countries.

- Most countries looked more closely at adopting measures to restrain energy consumption.

- Renewable energy sources were studied seriously (in some cases for the first time) to determine whether and where they could be used economically.

The thrust of these changes has continued into the new century. Throughout the world it was found possible to use significantly less energy per unit of economic activity. The use of oil for electricity production was greatly reduced and the use of natural gas increased. At the start of this century, wind is the fastest-growing source of electricity.

Continuing a trend predating the oil crisis, the demand for primary energy per unit of Gross Domestic Product (i.e. "energy intensity") has shown a significant decline (1.3% per year) in OECD countries, and this is expected to be the case also in developing countries in the future. However, at the same time the electricity consumption per unit of Gross Domestic Product has been growing steadily, reflecting a strong increase in the proportion of electricity used in all countries.

The role of electricity is increasing because it is an extremely versatile energy source which can be generated from a wide range of fuels and can easily be channelled to the point of use. Electricity generation uses some 40% of the world's total primary energy supply.

Electricity is uniquely useful for driving machinery and for lighting in both industry and homes. However, it is also used for heating and in other ways for which alternatives are readily available. It can be argued that in view of the relatively low efficiency of energy conversion to electricity (typically around 35%) alternatives such as natural gas should be used wherever possible for heating (at double the efficiency)[3]. Conversely, it can be argued that uranium and coal resources are large relative to gas resources, that the most abundant primary fuel should be applied wherever possible, and that hence electricity use for heating (at almost 100% end use efficiency) is desirable if it comes from coal or nuclear power despite a much higher consumption of primary fuel.

In one sense the Sun is the world's most abundant energy source and the desirability of applying it more widely to direct heating and even, eventually, for large-scale generation of electricity hardly needs emphasis. Meanwhile wind is increasingly harnessed for electricity. Questions concerned with the production of electricity from renewable sources are discussed in section 2.4.

In the following chapters electricity demand, use and generation are the focus of discussion. In particular they discuss the use of nuclear energy to generate electricity. The main nuclear fuel concerned is uranium, a metal which at present has virtually no other civil uses.

1.6 FUTURE ENERGY DEMAND AND SUPPLY

Where will we obtain our future energy needs? There are a number of uncertainties:

- Oil production peaked in 1979 and did not return to that level until 1994. Production costs have increased little since 1973. Prices depend largely on political factors.

- Natural gas production, while increasing rapidly now, is likely to approach its peak in many countries in the next couple of decades.

- Underground coal is costly to mine, and all coal use gives rise to concern about its effect on global warming.

- There is limited scope for utilizing renewable energy resources.

- Further scope for energy conservation is limited without radical changes in lifestyle in developed countries, and is minimal in developing countries.

Until the early 1970s the world's energy supplies were easily and cheaply bolstered by oil and natural gas whenever consumption tended to exceed supply. After 1973, however, many industrialized nations set out to develop other strategies including greater use of nuclear energy. Looking ahead, it is not just the industrialized countries which will dominate the scene. By 2030 China is projected to be using almost as much oil as the USA does now[4]. World energy consumption has been rising steadily for many decades. As economic growth occurs in most nations, increase in energy demand is an inescapable part of this growth. Also, growth of the world's population is expected to continue towards 8 billion by 2025, further increasing the demand for energy. Fossil fuels account for 90% of the projected growth in energy demand to 2030.

---

[3] Considering the whole sequence from production to end use, the efficiency of gas and oil for heating is often about 40-45%. For modern high efficiency gas furnaces, the value increases to about 70%, but overall depends on distance from the gas sources.
[4] Projections in this chapter are from IEA 2004 *World Energy Outlook*.

Today, oil demand continues to increase, while available resources decline. Production has exceeded discoveries since the 1980s, and despite very intensive exploration effort, consumption is now twice the rate of discovery. Most conventional reserves are in geopolitically uncertain parts of the world, and are difficult to access. A lot of newly-discovered oil requires greater effort in refining. Other reserves, such as tar sands, pose major problems to develop on any large scale.

Natural gas is less constrained, but again most reserves are located in geopolitically uncertain areas, and transport becomes a major problem. Moving it as liquefied natural gas (LNG) consumes up to 30% of it. Most of the projected increase in OECD demand to 2030 comes from the power sector.

Coal remains very abundant and tends to be located closer to where it may be used. It is economically attractive to use on a large scale, but delivers the greatest contribution to greenhouse gas emissions of any fossil fuel. Projected increase in demand to 2030 is for the power sector.

Renewable energy sources cannot meet the extent of the demand. Cost and the diffuse and intermittent nature of these sources (apart from hydro) limit their potential.

One third of the world's population does not have access to electricity supply, and a further third does not enjoy reliable supply. There is a huge need to address these shortcomings and expectations in the context of the overall implementation of sustainable development principles and reduction of poverty.

There is also a rapidly increasing demand for potable water in many developing areas (e.g. North Africa and the Arab Gulf States) that must be satisfied by desalinating facilities (cf. Chapter 6). This will further increase energy demand.

Future energy growth rate worldwide is projected to average 1.7% per year to 2030[5]. Achieving even this level of annual growth will require both some expansion of known supply and continuing efforts in energy conservation to increase the efficiency of energy use.

Figure 3: Electricity demand

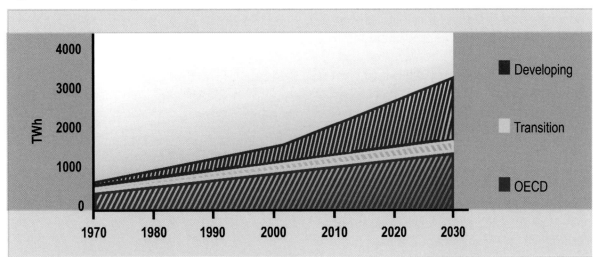

Source: OECD/IEA World Energy Outlook 2004.

[5] OECD/IEA 2004.

Since the 1970s economic factors have constrained energy demand and have resulted in unprecedented increases in energy efficiency in industry and transport, at least in the OECD countries. Primary energy consumption in OECD is forecast to increase only marginally, while that in developing countries is expected to grow very much faster.

Electricity demand is growing much faster than overall energy demand. Where world total energy demand is expected to increase 60% from 2002 to 2030, electricity demand is expected to double between 2002 and 2030[6], with most of the growth in developing countries (compare Figures 2 and 3). World electricity demand is projected to be about 31,600 billion kWh (TWh) in 2030, compared with 16,000 billion kWh in 2002, with the largest increase – almost 4000 billion kWh – being in China.

The future scope for energy conservation depends on the sector involved. Where energy is a significant input to industrial processes or to transport, or a major cost to consumers such as with motor vehicles, major steps have already been taken to increase efficiency and hence lower costs. But where energy costs are relatively less significant, such as in commercial and residential buildings, there is likely to be much greater scope for improvement.

Energy conservation is very difficult to project. To continue to be effective, it requires a present response to future prospects of higher energy costs. It demands an attitude to energy use and lifestyle which is increasingly conservation-oriented, so that the rate of increase in overall energy consumption remains depressed after the initial easy fixes have been achieved. Despite popular acceptance of environmental ideas, there is

little evidence of such an attitude taking precedence over comfort and amenity anywhere in the world.

---

[6] OECD/IEA 2004 *World Energy Outlook.*

Table 3: Energy conversion: the heat values and carbon coefficients of various fuels

	heat value		% carbon	CO_2
Hydrogen	121	MJ/kg	0	0
Petrol/gasoline	44-46	MJ/kg		
Crude oil	45-46	MJ/kg	89	70-73 g/MJ
	37-39	MJ/L		
Methanol	22	MJ/kg		
Liquefied petroleum gas (LPG)	49	MJ/kg	81	59 g/MJ
Natural gas (UK, USA, Australia)	38-39	MJ/m^3	76	51 g/MJ
(Canada)	37	MJ/m^3		
(Russia)	34	MJ/m^3		
as LNG[7] (Australia)	55	MJ/kg		
Hard black coal (IEA definition)	>23.9	MJ/kg		
(Australia and Canada)	24-30	MJ/kg	67	90 g/MJ
Sub-bituminous coal (IEA definition)	17.4-23.9	MJ/kg		
(South & West Australia)	13.5-19.5	MJ/kg		
Lignite/brown coal (IEA definition)	<17.4	MJ/kg		
(Australian average)	9.7	MJ/kg	25	
(Loy Yang, Australia)	8.15	MJ/kg		1.25 kg/kWh
Firewood (dry)	16	MJ/kg	42	94 g/MJ
Natural uranium, in LWR[8]	500	GJ/kg	-	-
in LWR with U & Pu recycle	650	GJ/kg	-	-
in FBR[9]	28,000	GJ/kg	-	-
Uranium enriched to 3.5%, in LWR	3900	GJ/kg	-	-

Sources: OECD/IEA Electricity Information 2004, for coal.

Australian Energy Consumption and Production, historical trends and projections, ABARE Research Report 1999.

Uranium figures are based on 45,000 MWD/t burnup of 3.5% enriched U in LWR.

(MJ = 10^6 Joule, GJ = 10^9 J, % carbon is by mass, g/MJ=t/TJ, C to CO_2: x 3.667)

MJ to kWh @ 33% efficiency: x 0.0926

toe = 41.868 GJ

[7] LNG = liquefied natural gas
[8] LWR = light water reactor
[9] FBR = fast breeder reactor

2

ELECTRICITY TODAY AND TOMORROW

2.1 ELECTRICITY DEMAND

Electricity demand in an industrial society arises from a number of sources, including:

Industry:
- Some running on a 24-hour basis
- Some working 8-10 hours only on weekdays

Commerce:
- Most working 10-15 hours per day

Public transport:
- Running during day and evening

Domestic, homes:
- Heating mostly during day and evening, (seasonal)
- Cooling (seasonal)
- Cooking morning and evening
- Off-peak water and space heating, especially during the night (in some systems)

It is clear from the above list why electricity demand fluctuates throughout every 24-hour period, as well as through the week and seasonally. It also varies from place to place and from country to country, depending on the mix of demand, the climate, and other factors. A daily load curve for an electricity system in a temperate climate is shown in Figure 4. From this it can be seen that there is a base load of about 60% of the maximum load for a weekday. This load curve is typical for developed countries.

The base-load demand for continuous, reliable supply of electricity on a large scale is the key factor in any system. The main investment of any electric utility is to meet that kind of demand.

As well as the daily and weekly variations in demand, there are gradual changes occurring in the pattern of electricity demand from year to year. In projecting demand patterns a decade or more into the future, planners must take note of such factors as:

- The changing pattern of seasonal peak demands; for example as summer air conditioning becomes more common.
- The impact of increased electrification of public transport.
- The possible electrification of private transport, either directly or through the use of hydrogen (produced by electrolysis) in fuel cells.
- The effect on supply systems of increasing use of solar water heating with electrical boosting during periods of adverse weather.
- The effect of incentives to increase off-peak electricity demand (and minimize peak demand) for water and space heating.
- The practical effect of energy conservation measures, such as insulation and more energy-efficient building and appliance design.
- The role of renewable energy sources providing electricity when they can, and political coercion on utilities to buy or supply that electricity preferentially at higher cost than other sources.
- Any increase in other dispersed electricity generation.
- Industry needs and how they are changing.
- Improvements in the ability to transmit electricity long distances; for example, 50 years ago 600 km was the maximum distance for efficient transmission; in the 1960s new technologies enabled transmission over 2000 km, and today it is greater still.

Looking further ahead, there is major scope for the use of base-load electricity to charge the batteries for personal motor vehicles. In the last couple of years the popularity of hybrid cars, such as the Toyota Prius, and also Honda's slightly different approach with hybrid diesel vehicles have put us within reach of practical electrical motoring for many people. This development has been enabled by the advent of much more efficient battery technology[1],

---

[1] The Toyota Prius nickel hydride battery in 2005 could deliver 21 kW from a mass of 45 kg. It held 6.6 amp hours at 201 volts (1.3 kWh) and had an 8 year/160,000 km warranty.

and with a further increase in battery capacity the possibility of using mainly energy derived from off-peak power, charging when parked overnight, is enhanced. This will mean that there is less reliance on the on-board internal combustion motor and more reliance on base-load power.

Some of these factors will affect total electricity consumption, while others will influence the relative importance of base-load demand. Production economics will require that as much of the electricity as possible is supplied from base-load generating plant. Government policies in many countries create scope for occasional input from any renewable generating capacity linked to the system.

2.2 ELECTRICITY SUPPLY

Because of the large fluctuations in demand over the course of the day, it is normal to have several types of power stations broadly categorized as base-load, intermediate-load or peak-load stations.

The base-load stations are usually steam-driven and run more or less continuously at near rated power output. Coal and nuclear power are the main energy sources used.

Intermediate-load and peak-load stations must be capable of being brought on line and shut down quickly once or twice daily. A variety of techniques are used for intermediate- and peak-load generation, including gas turbines, gas- and oil-fired steam boilers and hydroelectric generation.

Peak-load equipment tends to be characterized by low capital cost, and its relatively high fuel cost (unless hydro) is not a great problem.

Base-load plant is designed to minimize fuel cost, and the relatively high capital cost can be written off over the large amounts of electricity produced continuously.

Lowest overall power costs to the consumer are obtained when the peak-load increment is

Figure 4: Load curve of an electricity system

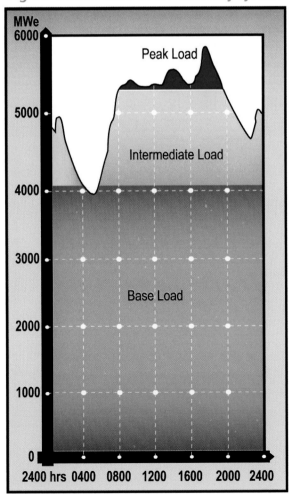

Source: Victorian Power Exchange (VPX).

Load curve of the electricity system in Victoria, Australia, through one winter weekday in June 1996 showing the relative contributions of base-, intermediate- and peak-load plant duty. The shape of such a curve will vary markedly according to the kind of demand. Here, the peaks reflect domestic demand related to a normal working day, with household electric hot water systems evident overnight.

Note that the base load here is about 4100 MWe, and while total capacity must allow for at least 50% more than this, most of the difference can be supplied by large intermediate-load gas-fired plant, or to some extent by adjusting the output of the base-load plant. The peak loads are typically supplied by hydro and gas turbines. In the wholesale electricity market, power stations bid into the market and compete for their energy to be despatched, so economic factors tend to determine the sources of supply at any particular moment.

very small and a steady base load utilizes most of the available generating capacity fairly constantly. Any practical system has to allow for some of the plant being unserviceable or under maintenance for part of the time. Installed capacity should therefore be about 20% more than maximum load in a system, providing a reserve.

Base-load plants are likely to make up over half of a system's total generating capacity, and produce more than 85% of the total electrical energy (cf. Figure 4). Almost one third of such a system's capacity can broadly be classified as intermediate-load plant, supplying power throughout the working day and evening. The balance is peak-load in the strict sense, supplying short-term energy demand during high-load periods of the day or in emergencies, and with unit power cost being less critical.

The capital cost of peak-load equipment, such as gas turbines, is about half that of base-load coal-fired plant, and in addition it can be installed much more quickly. However, the fuel cost is relatively high compared with coal in a base-load station, per unit of power generated. Modern combined cycle gas turbine facilities, which have efficiencies substantially greater than that of coal-fired plants, reduce the difference.

Pumped water storage, using available base-load capacity overnight and on weekends, may be developed where topography permits, as an alternative to peak-load thermal power stations[2]. The capital cost may be low where

there is existing hydroplant and such installations will have the effect of increasing the extent to which base-load equipment can contribute to total load through the week. However, there is an efficiency loss relative to inputs of around 35%.

In future the base-load contribution may be increased by using the surplus power in non-peak periods to make hydrogen, either for peak generation or for transport fuel (see section 6.1). A further means of increasing the utilization of base-load plant is enabling it to follow the load to some extent, by varying the output.

As in other industries, there are economies of scale. Larger steam units result in reduced capital cost per kilowatt capacity, especially for base-load equipment. This means that location is sometimes determined as much by the supply of cooling water as by the fuel source. However, large power station units require a substantial electrical transmission grid and overall generating system to enable them to be operated effectively. Hence there are many situations where the economic virtues of small-scale gas-fired generating plants are put forward.

Photograph supplied by Vattenfall

Goldisthal in Thuringia, Germany's largest pumped water storage plant

[2] See also later part of section 2.5.

2.3 FUELS FOR ELECTRICITY GENERATION TODAY

This book considers principally the question of electricity generation. In industrialized countries electricity generation takes about 40% of the primary energy supply. An increasing constraint on choosing the fuel for this is the carbon emissions involved.

In densely populated areas of the world, such as Japan and many parts of Europe and North America, coal supply is relatively remote from electricity demand. Also the high density of population and industrialization has limited the attractiveness of coal not only from a cost but also an environmental point of view (see Chapter 7). Therefore the desirable criteria for a fuel for base-load electricity generation in such parts of the world may be represented thus:

- It should be relatively **cheap**, giving low-cost power.
- Unless it can be supplied from a source very close to the power station, it should ideally be a **concentrated** source of energy, which can therefore be economically transported and readily stockpiled.
- It should have regard to the **scarcity of the resource** and **alternative valued applications** (such as burning directly, or chemical feedstock).
- **Wastes should be manageable**, so that they produce a minimum of **pollution** and environmental disturbance, including long-term global warming effect.
- Its use must be **safe** both in routine operation and during possible accident scenarios.

Of the three principal fuels available for base-load electricity generation, uranium often fits these criteria better overall than coal or gas, especially if the coal must be transported very far.

National energy strategies will vary according to the indigenous resources of each country, the economics of importing fuels (or electricity), the amount of industrialization and the security of supply.

An energy-rich country such as the USA has a variety of options. However, even in parts of the USA, transporting large quantities of coal long distances adds significantly to costs.

Japan lacks indigenous energy resources and relies almost entirely on imports. Oil was once the most convenient fuel import and the country depended on it for a large proportion of its energy needs, including electricity generation. Coal then became increasingly used for this purpose. Nuclear fuel has the advantage that so little is required and transport costs are negligible. Also, variations in the price of the fuel have very much less impact than with coal or gas.

Figure 5 shows how electricity is produced in certain countries and in Europe. In all countries the demand for electric power is increasing steadily (mostly 3% to 4% p.a.). The diagram shows that coal provides a lot of the primary energy input for electricity in the USA and Europe, while in Japan and Canada the proportion is less. Europe, USA, Japan and South Korea have about one fifth to one third of their electrical power being generated from nuclear reactors.

Even in this new century, a substantial amount of the fuel used in each country for generating electricity still consists of increasingly scarce and hence rather precious oil. This is most obvious and acute in Japan, though it has markedly reduced its dependence on oil for electricity in the last 30 years, and plans to increase the proportion of electricity generated from nuclear energy. Russia aims to increase the proportion of electricity generated by nuclear power explicitly to maximize gas exports to Europe. Both Russia and UK have a high dependence on gas.

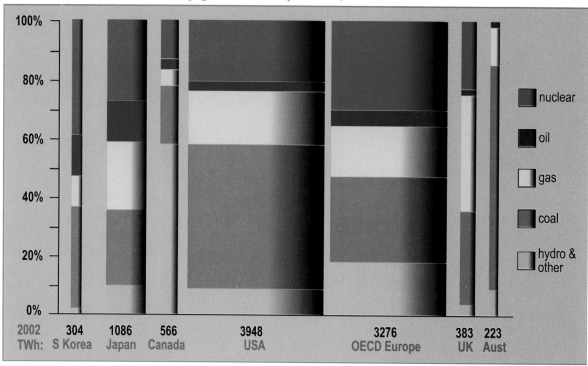

Width of each bar is indicative of power generated (gross production)
Source: OECD/IAEA 2003 "Electricity Information" 2002, Table 4.

2.4 PROVISION FOR FUTURE BASE-LOAD ELECTRICITY

In considering the future beyond a decade hence there are a couple of practical matters which cannot be overlooked. One is the **lead time** for constructing generating plant. A commitment today regarding a large base-load generating plant means that plant should be commissioned in five to ten years time. It can then be expected to have an operating life of up to 60 years. Thus today's investment decisions regarding electric plant cannot change the overall pattern of a country's generating system for several decades.

Even combined cycle gas turbines (CCGT), which can be put into service in less than two years from date of order, and which became very popular in the 1990s, cannot make a substantial short-term change to the overall energy supply situation. If we are considering new technologies for coal or nuclear base-load

plant, the lead time is longer. It follows that much of the technology in use today will inevitably be in use for several more decades – it cannot be quickly abandoned.

The other practical matter relates to size. In some things small is appropriate and, given low labour costs, also efficient. In mining fuels and generating electric power however, the economic constraints involved generally dictate that operations and plant be as large as practicable. Where the scale is reduced, the unit costs inexorably increase. With conventional types of plant, large-scale installations are inevitable in urbanized and industrialized nations, where much electricity demand is concentrated in small areas of the country.

These practical matters of long lead time and large-scale installations point to the need for careful assessment of future trends in electricity use to ensure that tomorrow's supply systems will effectively cope with tomorrow's electrical demand.

A major policy challenge is that deregulated electricity markets make the financing of any capital-intensive generating plant more difficult, even if its output is at lower cost than alternatives, and that issue has yet to be addressed in most countries. Furthermore, the technology used must be matched to the task. The big question facing governments, which determine policies, is identifying the most appropriate means of generating base-load electricity for a particular region in the future. What are the options?

Conservation:

One possibility may be to use less energy by practising rigorous conservation, principally through increased energy efficiency in use. This approach is relevant to many applications in developed countries, and can be applied to new installations in all countries. If the USA, the UK and Japan could each use less electricity, such a strategy might, by itself, eventually eliminate the oil-fired component in two of these countries and markedly reduce it in the third. Energy conservation in general is discussed in section 1.5. However, such conservation has a greater effect on total energy use than on actual electricity, and an increased proportion of electricity in the overall energy mix is often a prime means of conservation.

Oil:

In 2002 some 7% of all electricity came from burning oil – much less than a decade earlier, while in 1973 it accounted for a quarter of the world's power generation. Oil is uniquely important as the source of very portable and energy-rich petroleum products used for mobile transport; moreover, both oil and gas have important uses in the petrochemical industry as feedstock for the manufacture of plastics, fertilizers and pharmaceutical products. Burning oil in a steam-cycle power plant for base-load electricity generation where other fuels are economically available is questionable. Generally today oil is used for power generation in areas remote from natural gas resources and coalfields, in relatively small installations.

Natural Gas:

Gas today plays a major and steadily-increasing role in power generation (19% of world electricity in 2002). While gas prices have been low and gas turbines relatively cheap and quickly built, it has been a most attractive fuel. It has the distinction of giving rise to less carbon dioxide than coal, and hence is favoured in a short-term perspective to displace some coal for base-load power.

Natural gas is a superbly useful resource. It can be drawn from the earth, easily and economically distributed via large pipelines, then cheaply reticulated to small-scale points of use where it can be used as a fuel very efficiently (up to 90% at end use, allowing for flue losses). It can be liquefied for shipping overseas (for example as LNG to Japan, Korea and the USA). It is also a valuable chemical feedstock for manufacturing.

This means that large-scale use of gas for generating electricity, where less versatile alternatives are readily available, is likely to encounter acute economic constraints as gas prices rise – as in several parts of the world early this decade. There are also ethical questions, particularly relating to intergenerational equity. In short, our grandchildren may later wish that the current "dash for gas" had been more restrained, and had left more gas for them to use in higher-value applications.

Coal:

Of the fuels for base-load electricity generation, coal is at present the most important.

Coal plays the major role in most countries and has done so for many years, currently providing 39% of the world's electricity. Modern coal-fired power stations are more efficient than in the past, and at extra cost some of the environmental effects of burning high-sulphur coals can be eliminated, even if the global warming effect due to the production of massive amounts of carbon dioxide presently cannot (see Chapter 7). However, a lot of work is being done on "clean coal" technologies designed to reduce carbon emissions from coal burning. While these look technically feasible, the cost is likely to be very high, and they have yet to be deployed commercially.

Coal from large open cut mines is fairly cheaply obtained, but the costs of transport over long distances can make it less attractive than alternatives. If large quantities of coal are mined in one locality and shipped across a continent or overseas (for example, from Australia, Canada or South America to Japan or Europe), its handling and transport imposes significant costs and involves the consumption of further energy.

Also, like oil and gas, coal has important uses other than as a fuel. Carbon, even in steaming coal, is needed in large quantities for metal smelting, for future conversion to gas and liquid fuels, and for other purposes. Although reserves are large, conservation will become increasingly important.

Uranium:

The only other fuel which is a present option for base-load electricity is uranium. While relatively large amounts of ore may be mined and treated, two or three 200-litre drums of uranium oxide

Coal-fired heat and power plant located in Pruszków, Poland

Photograph supplied by Vattenfall

(U_3O_8) concentrate leaving the mine contain enough energy to keep large cities supplied with power for a day, so it is very concentrated and portable. It also has significant environmental advantages (see Chapter 7) which are increasingly recognized. Nuclear power is a mature technology – it is now half a century since the first commercial reactor came on line, and 60 years since nuclear fission (see Chapters 3 and 9) was first controlled.

In that time over 12,000 reactor-years of operating experience have been acquired with commercial reactors, and about the same from similar (but smaller) reactors in naval use.

Today there are some 440 nuclear power reactors in operation in 30 countries, including several developing nations. They provide about 16% of the world's total electricity.

Many more nuclear power stations are under construction or firmly planned. The reliability, safety and economic performance of nuclear power relative to coal or oil (see also section 2.6 and Chapter 7) has been demonstrated in many countries, especially those where at least one fifth of their electricity is generated by nuclear power. France generates three quarters of its electricity from nuclear power and is the world's largest electricity exporter.

Table 5 gives an indication of the different kinds of nuclear power reactors currently being used for electricity generation. In the longer term fast neutron reactors (see section 4.5) have the potential for vastly increasing the electric power yield from uranium, though resources are abundant.

Apart from military weapons and naval propulsion, uranium has no significant uses other than for electricity generation and for making medical and industrial isotopes. At least 95% of the world's uranium production today goes into electricity generation (the balance to naval propulsion and isotope production – see Chapter 6).

The potential of nuclear power for electricity generation, using uranium as a fuel, is principally applicable to nations which have large blocks of electricity demand. Today's nuclear power stations tend to be built in sizes from 500 megawatts electrical (MWe) to about 1600 MWe, anything smaller currently being less attractive economically. However, there are some developing nations which have moderate-sized electricity production and distribution systems and/or the need for cogeneration (for example, electricity and potable water production). These are able economically to use reactors in the 100 to 200 MWe size range where expensive oil-fired generation is the main alternative.

Nuclear Fusion:

Commercial nuclear fusion is still only a future hope. As well as looking for ways to harness incident sunlight, people have for a long time dreamed of taming the process which generates that light and heat – bringing the Sun right down to Earth. The process concerned is called nuclear fusion (as distinct from fission, see Chapter 3). The favoured method for achieving controlled fusion involves joining the nuclei of deuterium and tritium atoms (heavy isotopes of hydrogen) together at very high temperatures – about 100 million degrees Celsius. No method of sustaining such temperatures under stable conditions has yet been demonstrated. However, research continues, particularly in Japan, Europe, USA and Russia, and notably in the ITER facility being built in France. Perhaps some time in the next half-century heat from fusion will be harnessed to generate electricity. Fusion technology would be best suited to large-scale base-load applications, such as supplying cities and industrial regions.

The deuterium fuel is relatively abundant in seawater, but tritium is either derived from lithium, or produced in heavy water-moderated reactors. Almost limitless energy would be available if a deuterium-deuterium reaction could be achieved, but this requires much higher temperatures than the deuterium-tritium process. Controlled fusion of ordinary hydrogen nuclei as occurs in the Sun seems unlikely ever to be achieved on Earth, because the conditions required are even more extreme. The big advantage of all these reactions is that only small quantities of radioactive wastes are expected. Disadvantages include projected high cost, the high radioactivity created in structural components of the plant, the cost of producing tritium gas and the hazard of handling it.

2.5 RENEWABLE ENERGY SOURCES

Technology to utilize the forces of nature for doing work to supply human needs is as old as the first sailing vessel. There is a fundamental attractiveness about harnessing such forces in an age which is very conscious of the environmental effects of burning fossil fuels. They are virtuously clean compared with fossil fuels. Consequently a huge amount of effort, R&D and investment has gone into them in recent years.

Sun, wind, waves, rivers, tides and the heat from radioactive decay in the earth's crust as well as biomass are all abundant and ongoing, hence the term "**renewables**". Only one, the power of falling water in rivers, has been significantly tapped for electricity so far – 16% of world generated power is hydroelectric, though tidal flows and wind may perhaps one day catch up. Solar energy's main human application has been in agriculture and forestry, via photosynthesis, and increasingly it is harnessed for heat. Biomass (e.g. sugar cane residue) is burned where it can be utilized, and government policies will see it increase in OECD countries. Natural geothermal power generation, tapping underground steam, is important in a few localities, and hot fractured rock (HFR) geothermal[3] shows promise in others. The others are little used today.

Tidal flow turbines would seem to have greater potential than wind to deliver power more or less continuously, but they have yet to be proven commercially.

Turning to the use of less predictable **intermittent renewable energy sources for electricity**, there are immediate challenges in actually harnessing them. Apart from photovoltaic (PV) systems, the question is how to make them turn dynamos to generate the electricity. If it is heat which is harnessed, this is via a steam generating system.

If the fundamental opportunity of renewables is their abundance and relatively widespread occurrence, the fundamental problem, especially for electricity supply, is the variable and diffuse nature of solar and wind energy.

This means either that there must be reliable duplicate or back-up sources of electricity, or some means of electricity storage on a large scale. Apart from pumped-storage hydro systems, no such means exist at present and nor are any in sight. For a stand-alone system the energy storage problem remains paramount. Any substantial use of solar or wind for electricity in a grid means that there must be allowance for near 100% backup with hydro or fossil fuel capacity. This gives rise to very high generating costs by present standards, but in some places it may be the shape of the future.

There are now many thousands of wind turbines operating in various parts of the world, with a total capacity of 59,000 MWe at the end of 2005. This has been the most rapidly growing means of electricity generation in the last decade and provides a valuable complement to large-scale base-load power stations. Where there is an economic backup which can be called upon at very short notice (e.g. hydro), a significant proportion of electricity can be provided from wind. The most economical and practical size of commercial wind turbines is now up to 2 MWe, grouped into wind farms up to 200 MWe. Some new turbines are 5 MWe. Depending on site, most turbines operate at about 25% load factor over the course of a year (European average), but some reach 33%. Wind is projected to supply 3% of world electricity in 2030, and

---

[3] HFR involves pumping water down and through hot rocks, or using hot brine from deep granites some 4-5 km underground. These rocks are hot - around 250°C - because they have high levels of radioactivity and are insulated. They typically have 15-40 ppm uranium and/or thorium, but may be ten times this.

perhaps 10% in OECD Europe.

Intermittent renewable sources cannot be controlled to provide either continuous base-load power, or peak-load power when it is needed. In practical terms they are therefore limited to some 10% – possibly 20% if coupled with nearby hydro – of the capacity of an electricity grid. They cannot directly be applied as economic substitutes for fossil fuels or nuclear power, however important they may become in particular areas with favourable conditions. Nevertheless, such technologies can and will contribute helpfully, especially where there is political will to make consumers subsidize them, even if they are unsuitable for carrying the main burden of supply.

Photograph supplied by Vattenfall (Hans Blomberg)

Olsvenne 2, Sweden's largest wind power plant

If there were some way that large amounts of electricity from intermittent producers such as solar and wind could be stored efficiently, the contribution of these technologies to supplying base-load energy demand would be much greater.

Already in some places pumped storage is used to even out the daily generating load by pumping water to a high storage dam during off-peak hours and weekends, using the excess base-load capacity from coal or nuclear sources. During peak hours this water can be used for hydroelectric generation. Relatively few places have scope for pumped storage dams close to where the power is needed, and overall efficiency is low (around 65%). Means of storing large amounts of electricity as such in giant batteries or by other means have not been developed.

Environmental aspects of renewables:

Renewable energy sources have a completely different set of environmental costs and benefits to fossil fuel or even nuclear generating capacity.

On the positive side they emit no carbon dioxide or other air pollutants (beyond some decay products such as methane from new hydro-electric reservoirs), but because they are harnessing relatively low-intensity energy, their "footprint" – the area taken up by them – is necessarily much larger. Whether large areas near cities dedicated to solar collectors will be acceptable, if such proposals are ever made, remains to be seen. Beyond using rooves, 1000 MWe of solar capacity would require at least 20 square kilometres of collectors, shading a lot of country to the extent that agricultural productivity would be minimal.

In Europe, wind turbines have not endeared themselves to neighbours on aesthetic, noise or nature conservation grounds, and this has arrested their onshore deployment, particularly in UK. At the same time, European non-fossil fuel obligations have led to the establishment of major offshore wind farms and the prospect of more.

However, much environmental impact can be reduced. Fixed solar collectors can double as noise barriers along highways, roof-tops are available already, and there are places where wind turbines would not obtrude unduly.

Figure 6: Fuel and waste comparison for uranium and coal

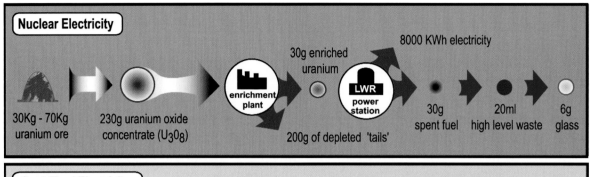

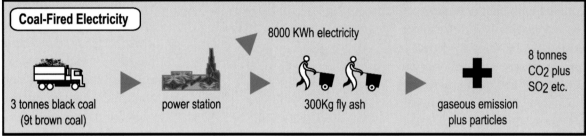

2.6 COAL AND URANIUM COMPARED

The only major fuel options for large-scale energy conversion to base-load electricity over the next several decades are coal and uranium.

Gas is an option in some places in the short term, but its great value as a direct fuel and the likelihood of significant price increases put the spotlight back on to coal and uranium. Choices between these alternatives will probably continue to depend principally on the final cost of electric power (including environmental costs), which varies significantly from site to site. Financing the capital expenditure involved may also be a factor.

Some general comparisons between coal and uranium as the principal fuels for base-load electricity generation are discussed in this section. Other comparisons which are principally environmental or related to health, which are the external costs, are discussed in more detail in Chapter 7.

Different quantities of materials are involved with energy conversion to electricity, starting with coal and uranium. In either case the amount of electricity considered is 8000 kWh, a conservative estimate of the amount required by one person in a developed country for one year.[4]

Using uranium as the fuel:

For 8000 kWh, between 30 kg and 70 kg of uranium ore from a typical Australian or older Canadian mine is needed to produce a handful (230 g) of uranium oxide concentrate. The uranium in this concentrate, is referred to as "natural uranium" and contains about 0.7% U-235, the fissile isotope of uranium. For most nuclear reactors the natural uranium is enriched in its U-235 isotope to yield about 30 grams of enriched uranium fuel (3.5% U-235, see section 4.2).

Irradiated fuel from light water reactors contains a useful quantity of fissile material and, in some countries, it is reprocessed to recover this. When light water reactor fuel is reprocessed, about 20 ml of liquid high-level

[4] The average consumption in industrialized countries is about 9000 kWh/yr (World Energy Council, 2000). Canadian consumption is 15,635 kWh/person/year, EU consumption is 5913 kWh/person/year and in the USA it is 12,640 kWh /person /year (OECD/IEA Electricity Information 2002).

waste remains. This then can be incorporated into less than 1 cm³ (6 g) of pyrex glass – about the size of a large coin and highly radioactive. Other wastes are also produced, but they are of much less significance (see section 5.1).

Using coal as the fuel:

About 3 t of high quality black coal (or 3.5 t of average black coal or 9 t of brown coal) can be fed into a power station to generate the same amount of electricity – 8000 kWh. This leaves a certain amount of ash, varying from a couple of barrow loads to half a tonne, depending on the particular coal used. Eight tonnes of carbon dioxide, which at atmospheric temperature and pressure would fill three full-sized Olympic pools (50 m x 15 m x 2 m), is produced. Depending on the coal, some sulphur dioxide (SO_2) is also produced. A common type of US coal might contain 2% to 3% sulphur, in which case possibly 100 kg of sulphur dioxide would require costly removal, or would add to the acid rain problems well known in the northern hemisphere. The environmental effects of these gaseous by-products of coal-fired electricity generation are considered in more detail in sections 7.1 and 7.2, and the costs of SO_2 removal are mentioned below. (Australian and Canadian coal generally contains less than 1% sulphur).

Years ago, most coal-fired power plants emitted more radioactivity than any nuclear plants of similar size! This was due to trace quantities of radioactive materials (e.g. up to 17 ppm U+Th in Australia and Canada) in the coal. With modern equipment this radioactivity is mostly retained with the fly ash and is buried with it.

2.7 ENERGY INPUTS TO GENERATE ELECTRICITY

Any electricity generation requires some energy inputs in mining, concentrating and transporting the fuel, manufacturing and constructing the plant, and dealing with the wastes. Energy use in mining and transport is closely related to quantities involved, and any comparison therefore favours uranium. On the other hand, the capital-intensive nature of the nuclear fuel cycle is reflected in the plant and the greater energy inputs to it.

The main energy input to the nuclear fuel cycle for reactors requiring enriched fuel may be in enriching uranium (see section 4.2). The following figures consider a 1000 MWe reactor run at 80% and therefore generating 7 billion kWh/yr. Conservatively, this would require about 195 t of natural uranium each year, which might be enriched to produce 27 t of uranium oxide fuel containing 24 t U at 3.5% U-235. After conversion to UF_6 this natural uranium would need 6 GWh of electricity to enrich it in a modern centrifuge plant (or up to 300 GWh in an older diffusion plant)[5].

Then there is fuel fabrication as well as construction and operation of the reactor to

Photograph supplied by Vattenfall

Digging for lignite coal in Germany, using an excavator

---

[5] At a tails assay of 0.25% U-235 in the enrichment plant, 5 SWU per kg of 3.7% enriched product is required, @ 50 kWh/SWU for the modern centrifuge plant or up to 2400 kWh/SWU for the older gaseous diffusion plant. The 195 t of natural uranium would leave the mine as 230 t U_3O_8.

include. The total energy inputs to the nuclear fuel cycle over a nuclear plant's full lifetime represent about 1.7% of the energy output[6]. Mining at Australia's Ranger (open cut) or Beverley (in situ leaching) mines uses energy equivalent to 0.05% of the mine's output, if that uranium is used in a light water reactor. If extremely low-grade uranium ore (0.01% U) is assumed, the life cycle energy input rises to about 3%.

Life cycle figures for coal range from 3.5% to 7% of the energy output required for inputs.

In mass terms the fuel inputs provide a stark contrast. Compared with uranium, about 20,000 times as much coal is required.

The energy payback time for a nuclear plant is about four months at full output.

See also: WNA information papers on "Energy Balances and CO_2 Implications" and "Energy Analysis of Power Systems".

2.8 ECONOMIC FACTORS

As well as comparing the quantities of fuel and wastes involved, the relative costs of different types of generating systems are important in considering options. This section focuses on the internal costs – those which need to be paid in the course of building and operating the plants. External costs are those which are actually incurred in relation to health and the environment but not paid directly by the electricity producer or consumer. These are large for fossil fuels, especially coal, and are considered further in Chapter 7. Equivalent costs for nuclear energy, notably waste management and disposal and decommissioning old reactors, are internalized and paid for by the consumers of their electricity.

A nuclear power station costs a lot more than a gas-fired station and somewhat more than a coal-fired station to build. But the nuclear fuel, including enrichment if needed, costs much less than oil, gas or coal. Hence the overall cost for energy conversion to electricity can come out much the same for nuclear as for coal-fired

Table 4: Comparative electricity generating cost projections for 2010 on

	nuclear	coal	gas
Finland	**2.76**	3.64	-
France	**2.54**	3.33	3.92
Germany	**2.86**	3.52	4.90
Switzerland	**2.88**	-	4.36
Netherlands	**3.58**	-	6.04
Czech Republic	**2.30**	2.94	4.97
Slovakia	**3.13**	4.78	5.59
Romania	**3.06**	4.55	-
Japan	**4.80**	4.95	5.21
Korea	2.34	**2.16**	4.65
USA	3.01	**2.71**	4.67
Canada	**2.60**	3.11	4.00

Cost in USA 2003 cents/kWh. Discount rate 5%, 40-year lifetime, 85% load factor.
Source: OECD 2005

[6] With older diffusion enrichment the figure could be up to 5%.

Source: US Utility Data Inst (pre 1995), Resource Data International (1995 -)

Note: The above data refer to fuel plus O & M costs only. They exclude capital since this varies greatly among utilities and states. Figures in Table 4 include capital.

plants. Table 4 quotes some comparisons for the projected costs of electricity compiled by the OECD, and Figure 7 shows the actual costs over more than a decade in the USA. Figure 8 shows the components of electricity cost for different means of generating it.

There are a number of US nuclear plants where capital costs overran during construction, and hence any normal calculation of their generating cost shows it to be very high. However, closing such plants would help neither owners nor customers, and in any case the criterion for running them is the cost of actual operation (Operation and Maintenance [O&M] plus fuel; see Figure 7). On this basis they compare favourably with coal and are cheaper than gas. Nearly 20 of these older US reactors changed hands over 1998-2005, and the escalating prices indicated the favourable economics involved. Regarding investment in new capacity, the capital costs are a major factor, and these are included in Table 4 and Figure 8.

In an earlier version of Table 4, OECD figures for plants starting operation in 2000 showed the importance of having coal near its point of use and using coal that is low in sulphur (which is expensive to remove from emissions). Costs

in the north eastern USA distinctly favoured nuclear energy, costs in the Midwest marginally favoured nuclear energy, and in the west, coal was often cheaper. This is reflected in today's distribution of nuclear plants. Having the location of electricity demand a long way from sources of cheap coal is the main reason for the steadily increasing use of nuclear power in many countries as compared with coal.

Actual electricity production costs in the USA (excluding capital) are shown in Figure 7. These are average figures including a lot of old coal and nuclear plant, and should be read with Figure 8.

Generally plant choice is likely to depend on a country's international economic situation. Nuclear power is very capital-intensive (see Figure 8), while fuel costs are relatively much more significant for systems based on fossil fuels. Therefore if a country such as Japan or France has to choose between importing large quantities of fuel or spending a lot of capital at home, simple costs may be less important than wider economic considerations.

Development of nuclear power, for instance, could provide work for local industries which build the plant and also minimize long-term commitments to buying fuels abroad. Overseas purchases over the lifetime of a new coal-fired plant in Japan, for example, may be subject to price rises which could be a more serious drain on foreign currency reserves than less costly uranium.

Uranium has the advantage of being a highly concentrated source of energy which is therefore easily and cheaply transportable. The quantity needed is very much less than for coal. One kilogram of natural uranium yields about 20 thousand times as much energy as the same amount of coal (see Table 3). In addition the fuel's contribution to the overall cost of electricity produced is relatively small, which means that even a large fuel price escalation will have relatively little effect.

However, as the long-term global environmental consequences of consuming fossil fuels, especially coal, create additional concern, the environmental advantages of nuclear power are also receiving more attention (see section 7.1) and will increasingly be reflected in the overall economics if costs are imposed on carbon emissions. Assigning carbon values to or imposing carbon costs on fossil fuel electricity generation changes the economic situation relative to nuclear energy. For instance, carbon values of $37 per tonne ($10 per tonne of CO_2) for typical coal will increase the electricity cost from that source by one cent per kilowatt hour but leave nuclear electricity costs unaffected.

See also: WNA information paper on Economics of Nuclear Power.

Although coal and uranium broadly compete for base-load electricity generation, most developed nations fortunate enough to have the option see a role for both.

As a general rule countries without cheap coal or plentiful gas tend to favour nuclear power as the lower cost option. In a few countries (e.g. Australia, where coal reserves and production potential far outweigh domestic needs) the use of coal for electricity generation is favoured over nuclear power. However, in a world perspective, the need for both is evident, and as electricity demand increases along with concern regarding possible global warming, a corresponding preference for nuclear power to generate base-load electricity seems inevitable.

Figure 8: Electricity generating cost in OECD 1990

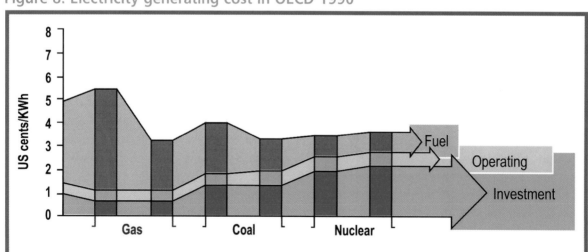

Source: OECD 1992, Electricity Supply in OECD, annex 9.
For different fuel costs (fossil fuels) or lead time (nuclear plants). Assumes 5% discount rate, 30-year life and 70% load factor.
Note: The key factor for fossil fuels is the high or low cost of fuels (top portion of bars), whereas nuclear power has a low proportion of fuel cost in total electricity cost, and the key factor is the short or long lead time in planning and construction, hence investment cost (bottom portion of bars). Increasing the load factor thus benefits nuclear power more than coal, and both these more than gas.

ELECTRICITY TODAY AND TOMORROW

36

3

NUCLEAR POWER

3.1 MASS TO ENERGY IN THE REACTOR CORE

While people until relatively recently must have thought they were converting mass to energy when they burned wood to cook meals and to keep warm, any student today would be aware that this was not the case. One form of carbon compound (the solid wood) was simply being converted to another (a colourless gas) which blew away. The hydrogen involved with the original compound also dispersed as water vapour. No measurable mass was lost, although energy was released. During the 20th century, as our understanding of nuclear physics developed, it was suggested that mass could in fact be turned into energy. This is what happens in a nuclear reactor, using atoms of particular metals such as uranium.

Uranium is 1.7 times more dense than lead, and is composed of atoms which have in their nucleus 92 protons (positively-charged) and about 140 neutrons (uncharged). One of the types of uranium atoms, or one of the uranium "isotopes" as they are called, has 143 neutrons. This uranium-235 (U-235) isotope is remarkable because when its nucleus is hit by a slow neutron (also known as a "thermal" neutron), the atom can split in two and release a lot of energy as heat. This is called nuclear "fission", and U-235 is thus a "fissile" isotope. In Einstein's terms some mass is lost and converted to energy. At the same time several fast neutrons are emitted from the split nucleus. If these are slowed by a moderator, such as graphite or water, they can cause other U-235 atoms to split, thus giving rise to a chain reaction (see also section 3.7 below and Figures 15 and 16).

The other main isotope of natural uranium, U-238, is not itself fissile in conventional reactors, but each atom can capture a neutron indirectly to become fissile plutonium-239. It is thus "fertile". Pu-239 behaves similarly to U-235 except that its neutron yield is slightly greater than that of U-235. About one third of the energy from a commercial nuclear reactor comes from fission of the plutonium-239 produced in the reactor.

The reactor core is loaded with uranium oxide fuel. In light water reactors this is enriched to 3.5% to 5% U-235 (see also section 4.2)[1]. It is typically in the form of ceramic pellets of UO_2 with melting point of about 2800°C, assembled inside zircaloy or stainless steel tubes and surrounded by coolant and moderator.

The moderator slows down the fast neutrons from the nuclear fission chain reaction so that they are more likely to cause ongoing fission. The slow neutrons cause further fission in U-235 atoms. If water is used as a moderator, the fuel must be enriched in the level of U-235 because of the tendency of water to also capture neutrons. If graphite or heavy water moderator is used, natural uranium will work as fuel.

Each such fission typically releases about 200 MeV, or 3.2×10^{-11} Joule (contrasting with about a million times less from chemical reactions such as combustion). Commercial nuclear power generation involves containing and controlling the fission reactions so that the heat can be used to make steam which in turn generates electricity. Removing the heat reliably is vital. Not only does the heat from fission need to be removed as the reactor operates, but for some time after shutting down, the decay heat from fission product radionuclides must also be removed from the core.

As fission takes place in the core the fuel changes. Its fissile content diminishes as "burnup" proceeds, and new elements – both fission products and transuranic elements – build up. Some of these are neutron absorbers, which progressively make the fuel less efficient. Compensation for this is provided by

[1] In CANDU reactors, natural uranium - 0.7% U-235 - is used.

progressively withdrawing control rods or reducing boron levels in the coolant, but at some stage – after about four years – the used fuel needs to be replaced. Typically one third of the fuel is replaced in each refuelling "outage".

The nuclear fuel cycle – the sequence of what is done to the fuel before it is used in the reactor and what happens to it afterwards – is described in section 4.2. A fuller account of the physics involved is in section 3.7.

3.2 NUCLEAR POWER REACTORS

In the middle of the last century an extraordinary variety of experimental nuclear reactors were built and operated, with every conceivable type of fuel, moderator and coolant. Gas-cooled graphite-moderated reactors were popular initially, but very soon the focus shifted to designs moderated by light water and using enriched uranium. These also predominated in naval use.

The most popular reactor design for generating electricity is the Pressurized Water Reactor (see Figure 9). In the core the uranium undergoes fission so that a lot of heat is released. The control rods shown regulate the rate of the reaction, and therefore the heat yield, by absorbing some of the moving neutrons. The core is surrounded by water and is enclosed in a very thick steel pressure

Figure 9: Pressurized water reactor (PWR)

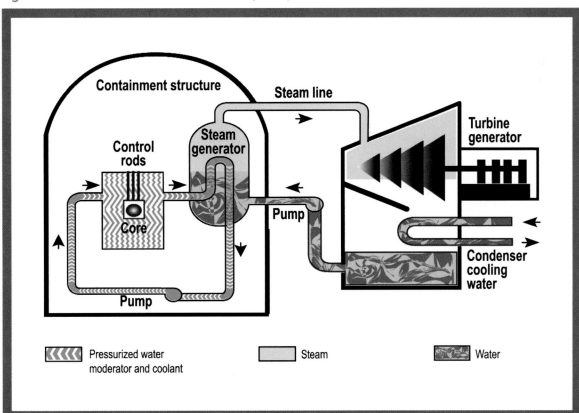

Table 5: Nuclear power plants commercially operable

Reactor type	Main countries	Number	GWe	Fuel	Coolant	Moderator
Pressurized water reactor (PWR)	USA, France, Japan, Russia	268	249	enriched UO_2	water	water
Boiling water reactor (BWR)	USA, Japan, Sweden	94	85	enriched UO_2	water	water
Pressurized heavy water reactor "CANDU" (PHWR)	Canada	40	22	natural UO_2	heavy water	heavy water
Gas-cooled reactor (Magnox & AGR)	UK	23	12	natural U (metal), enriched UO_2	CO_2	graphite
Light water graphite reactor (RBMK)	Russia	12	12	enriched UO_2	water	graphite
Fast neutron reactor (FBR)	Japan, France, Russia	4	1	PuO_2 and UO_2	liquid sodium	none
	TOTAL	441	381			

Source: Nuclear Engineering International Handbook 2005
GWe = capacity in thousands of megawatts.

vessel. The water, under high pressure, serves as both coolant and moderator. It is circulated to a heat exchanger (steam generator) where water in a separate circuit is turned into steam.

All this occurs in a big concrete or steel containment structure. The steam is fed to a turbine generator, much the same as those installed in coal-fired power stations. The uranium-fuelled core of a nuclear power reactor simply takes the place of a boiler or furnace burning coal (or other fossil fuel) to generate the steam.

Descriptions of the different kinds of reactors found in Table 5 may be found in the WNA information paper, *Nuclear Power Reactors*. An introduction to advanced reactor designs now coming on the market is in section 4.3.

Several components common to most types of nuclear reactors:

Fuel. Usually pellets of uranium oxide (UO_2) arranged in tubes to form fuel rods. The rods are arranged into fuel assemblies in the reactor core.

Moderator. This is material which slows down the neutrons released from fission so that they cause more fission. It is usually water but may be heavy water or graphite.

Control rods. These are made with neutron-absorbing material, such as cadmium, hafnium or boron, and are inserted or withdrawn from the core to control the rate of reaction or to halt it. (Secondary control systems involve other neutron absorbers – usually boron in the coolant, the concentration of which can be adjusted over time as the fuel burns up.)

Coolant. A liquid or gas circulating through the core so as to transfer the heat from it. In light water reactors the moderator functions also as coolant.

Pressure vessel or pressure tubes. Usually a robust steel vessel containing the reactor core and moderator/coolant, but it may be a series of tubes holding the fuel and conveying the coolant through the surrounding moderator.

Steam generator. Part of the cooling system where the heat from the reactor is used to make steam for the turbine.

Containment. The structure around the reactor core which is designed to protect it from outside intrusion and to protect those outside from the effects of radiation in case of any malfunction inside. It is typically a metre-thick concrete and steel structure.

Nuclear electricity output has been increasing.

Table 6: Nuclear power's role in electricity production

	NUCLEAR GENERATION 2005		OPERABLE at May 2006		CONSTRUCTION at May 2006		PLANNED at May 2006		PROPOSED at May 2006	
	TWh	% e	No.	MWe	No.	MWe	No.	MWe	No.	MWe
Argentina	6.4	6.9	2	935	1	692	0	0	-	-
Armenia	2.5	43	1	376	0	0	0	0	1	1000
Belgium	44.3	56	7	5728	0	0	0	0	-	-
Brazil	9.9	2.5	2	1901	0	0	1	1245	-	-
Bulgaria	17.3	44	4	2722	0	0	2	1900	-	-
Canada	86.8	15	18	12,595	0	0	2	1540	-	-
China - mainland	50.3	2.2	10	7587	5	4170	9	4600	19	15,000
Taiwan	38.4	20	6	4884	2	2600	0	0	-	-
Czech Rep.	23.3	31	6	3472	0	0	0	0	2	1900
Egypt	0	0	0	0	0	0	0	0	1	600
Finland	22.3	33	4	2676	1	1600	0	0	-	-
France	430.9	79	59	63,473	0	0	1	1630	-	-
Germany	154.6	31	17	20,303	0	0	0	0	-	-
Hungary	13	37	4	1755	0	0	0	0	-	-
India	15.7	2.8	15	2993	8	3638	0	0	24	13,160
Indonesia	0	0	0	0	0	0	0	0	4	4000
Iran	0	0	0	0	1	915	2	1900	3	2850
Israel	0	0	0	0	0	0	0	0	1	1200
Japan	280.7	29	55	47,700	1	866	12	14,782	-	-
Korea DPR (N)	0	0	0	0	1	950	1	950	-	-
Korea RO (S)	139.3	45	20	16,840	0	0	8	9200	-	-
Lithuania	10.3	70	1	1185	0	0	0	0	1	1000
Mexico	10.8	5	2	1310	0	0	0	0	2	2000
Netherlands	3.8	3.9	1	452	0	0	0	0	-	-
Pakistan	2.4	2.8	2	425	1	300	0	0	2	1200
Romania	5.1	8.6	1	655	1	655	0	0	3	1995
Russia	137.3	16	31	21,743	4	3600	1	925	8	9375
Slovakia	16.3	56	6	2472	0	0	0	0	2	840
Slovenia	5.6	42	1	676	0	0	0	0	-	-
South Africa	12.2	5.5	2	1842	0	0	1	165	24	4000
Spain	54.7	20	8	7442	0	0	0	0	-	-
Sweden	69.5	45	10	8938	0	0	0	0	-	-
Switzerland	22.1	32	5	3220	0	0	0	0	-	-
Turkey	0	0	0	0	0	0	0	0	3	4500
Ukraine	83.3	49	15	13,168	0	0	2	1900	-	-
UK	75.2	20	23	11,852	0	0	0	0	-	-
USA	780.5	19	103	98,054	1	1065	0	0	13	17,000
Vietnam	0	0	0	0	0	0	0	0	2	2000
WORLD	2626	16	441	369,374	27	21,051	38	40,737	115	83,620

Source: Reactor data: WNA to 31/5/06

Uranium: WNA reference scenario = 77,218 t U$_3$O$_8$, % e = % of total electricity from nuclear, TWh = billion kWh (source: IAEA 5/06), MWe: net

Operating = Connected to the grid; Construction = First concrete for reactor poured, or major refurbishment underway; Planned = Approvals & funding in place, or construction well advanced but suspended indefinitely; Proposed = Clear intention but still without funding and/or approvals. Canadian "planned" figure is for 2 laid up Bruce A reactors

In 2003 nuclear electricity generation was 2618 billion kilowatt hours, more than all electricity generated worldwide in the early years of the 1960s, and an increase of 17.5% over the previous ten years. The reasons for the overall growth are several.

Firstly, and most obviously, capacity is steadily increasing as new reactors come on line, as suggested by Table 6. At the end of 2005 there were 441 nuclear power reactors with a capacity of over 368,000 MWe operating in 30 countries, with 25 nuclear power reactors (20 GWe) under construction in 10 countries and 39 more units firmly planned. New reactor start-ups are partly offset by the closure of old plants, most of them smaller than those starting up.

Secondly, increased nuclear capacity in some countries is resulting from the uprating of existing plants. Power reactors in the USA, Belgium, Sweden, Spain, Switzerland and Germany, for example, have had their generating capacity increased. In 2004 alone such uprates totalled 153 MWe.

Thirdly, capacity or load factors are improving everywhere, so that more kilowatt hours come from the installed capacity. More than two thirds of the nuclear plants in the last few years have had load factors over 75%, up from average 67% average load factor in 1992. The average for the two main light water types is now over 80%. Many countries average over 80% load factor including the US, where nuclear power plant performance, at over near 90%, has moved into the top bracket. Recently the annual improvement in US reactor performance was equivalent to putting two to three large new power station units on line each year. To put it another way, the US increase from 65% load factor in the 1980s to almost 90% today is equivalent to adding 23,000 MWe capacity.

Fourthly, plant lives are being extended. Most nuclear power plants originally had a nominal design lifetime of 30 to 40 years, but engineering assessments have established that many plants can operate longer. Extending reactor operating life by replacing major components is often an attractive and cost-effective option for utilities. In the USA and Japan most reactors had confirmed lifespans of 40 years, but many – more than one third of those in the USA – have now been cleared to operate for 60 years. When the oldest commercial nuclear power stations in the world, Calder Hall and Chapelcross in the UK, were built in the 1950s, it was assumed that they would have a useful lifetime of 20 years. They were then authorized to operate for 50 years, though in fact they closed earlier for economic reasons.

New reactor start-ups seem likely to exceed the decommissioning of old reactors for several years at least, though most of the new reactors will be in Asia.

3.3 URANIUM AVAILABILITY

Uranium is ubiquitous on the earth.

Uranium is a metal approximately as common as tin or zinc, and it is a constituent of most rocks and even of the sea. Some typical concentrations are (ppm = parts per million):

High-grade orebody	2% U	20,000 ppm U
Low-grade orebody	0.1% U	1000 ppm U
Granite		4 ppm U
Sedimentary rock		2 ppm U
Average in Earth's continental crust		2.8 ppm U
Seawater		0.003 ppm U

An orebody is, by definition, an occurrence of mineralization from which the metal is economically recoverable. It is therefore relative to both costs of extraction and market prices. At present neither the oceans nor any granites are orebodies, but conceivably either could become so if prices were to rise sufficiently.

Measured "resources" of uranium, the amount known to be economically recoverable from orebodies, are thus also relative to costs and prices. They are also dependent on the intensity of exploration **effort, which has been low from the early 1980s to 2005.** They are basically a statement about what is known, rather than what is there in the Earth's crust.

Changes in costs or prices, or further exploration, may alter measured resource figures markedly. Thus any predictions of the future availability of any mineral, including uranium, which are based on current cost and price data and current geological knowledge, are likely to be extremely conservative.

From time to time concerns are raised that the known resources might be insufficient when judged as a multiple of present rate of use. But this is the "Limits to Growth" fallacy, a major intellectual blunder recycled from the 1970s, which takes no account of the very limited knowledge we have at any time of what is actually in the Earth's crust. Our knowledge of geology is such that we can be confident that identified resources of metal minerals are a small fraction of what is there.

With those major qualifications Table 7 gives some idea of our present understanding of uranium resources. It can be seen that Australia has a substantial part (about 30%) of the world's low-cost uranium, Kazakhstan 17%, and Canada 12%.

Table 7: World uranium resources

Known recoverable resources of uranium		
	tonnes U	**% of world**
Australia	**1,074,000**	**30%**
Kazakhstan	622,000	17%
Canada	439,000	12%
South Africa	298,000	8%
Namibia	213,000	6%
Russian Fed.	158,000	4%
Brazil	143,000	4%
USA	102,000	3%
Uzbekistan	93,000	3%
World total	**3,622,000**	

Reasonably Assured Resources plus Estimated Additional Resources – category 1, to US$ 80/kg U, 1/1/03, from OECD NEA & IAEA, Uranium 2003: Resources, Production and Demand, updated 2005 by Geoscience Aust.

Presently-known resources of uranium are enough to last for half a century – considering only the lower cost category, and with it used only in conventional reactors. This represents a higher level of assured resources than is normal for most minerals. Further exploration and higher prices will certainly, on the basis of present geological knowledge, yield further resources as present ones are used up. A doubling of price from present contract levels could be expected to create about a tenfold increase in measured resources, over time.

Widespread use of the fast breeder reactor (see section 4.2) could increase the utilization of uranium sixty-fold or more. This type of reactor can be started up on plutonium derived from conventional reactors and operated in closed circuit with its reprocessing plant. Such a reactor, supplied with natural uranium for its "fertile blanket", very quickly reaches the stage where each tonne of ore yields 60 times more energy than in a conventional reactor.

See also: WNA information paper on Uranium supply.

Reactor Fuel Requirements

The world's power reactors, with combined capacity of 368 GWe, require some 80,000 tonnes of uranium oxide concentrate from mines (or stockpiles) each year. While this capacity is being run more productively, with higher capacity factors and reactor power levels, the uranium fuel requirement is

"Yellowcake", the penultimate uranium compound in U3O8 production

increasing but not necessarily at the same rate. The factors increasing fuel demand are offset by a trend for higher burnup of fuel and other efficiencies, so demand is steady. (Over the 18 years to 1993 the electricity generated by nuclear power increased 5.5-fold while uranium used increased only just over 3-fold.) It is likely that the annual uranium demand will grow only slightly to 2010.

Fuel burnup is measured in MW days per tonne U (MWd/t), and many operators are increasing the initial enrichment of their fuel (e.g. from 3.3% to 4.0% U-235) and then burning it longer or harder to leave only 0.5% U-235 in the fuel. This might mean that typical burnup is increased from 33,000 MWd/t to 45,000 MWd/t.

Reprocessing of spent fuel from conventional light water reactors (see section 5.2) also utilizes present resources more efficiently, by a factor of up to 1.3 overall. At present the (reactor-grade) plutonium arising from reprocessing is used in fresh mixed oxide fuel (MOX), with depleted uranium from enrichment plants.

The net result from all this is a small reduction in the amount of uranium required ex-mine to fuel each kilowatt-hour produced.

3.4 NUCLEAR WEAPONS AS A SOURCE OF FUEL

An important source of nuclear fuel is the world's nuclear weapons stockpiles.

Since 1987 the USA and countries of the former USSR have signed a series of disarmament treaties to reduce the nuclear arsenals of the signatory countries by approximately 80%.

The weapons contain a great deal of uranium enriched to over 90% U-235 (i.e. about 25 times the proportion in most reactor fuel). Some weapons have plutonium-239, which can be used in diluted form in either conventional or fast breeder reactors.

Uranium

The surplus of weapons-grade highly enriched uranium (HEU) has led to an agreement between the USA and Russia for the HEU from Russian warheads and military stockpiles to be diluted for delivery to the USA and then used in civil nuclear reactors. Under the "megatons to megawatts" deal signed in 1994, the US government is purchasing 500 tonnes of weapons-grade HEU over 20 years from Russia for dilution and sale to electric utilities, for US$ 12 billion. This acquisition reached its halfway point in 2005 with the claim that this eliminated 10,000 nuclear warheads.

Weapons-grade HEU is enriched to over 90% U-235 while light water civilian reactor fuel is usually enriched to about 3% to 4%. To be used in most commercial nuclear reactors, military HEU must therefore be diluted about 25:1 by blending with depleted uranium (mostly U-238), natural uranium (0.7% U-235), or partially enriched uranium. The contracted HEU is being

blended down to 4.4% U-235 in Russia, using 1.5% U-235 (enriched tails). The 500 tonnes of weapons HEU is resulting in just over 15,000 tonnes of low-enriched (4.4%) uranium over the 20 years. This is equivalent to about 153,000 tonnes of natural uranium, more than twice annual world demand.

The purchase and blending down is being done progressively. Since 2000 the dilution of 30 tonnes per year of military HEU is displacing about 10,600 tonnes of uranium oxide mine production per year, representing about 13% of the world's reactor requirements.

In addition, the US Government has declared 174 tonnes of highly-enriched uranium (of various enrichments) to be surplus from its military stockpiles, and this is being blended down to about 4300 tonnes of reactor fuel. In the short term most of the military uranium is likely to be blended down to 20% U-235, then stored. In this form it is not useable for weapons.

Plutonium

Disarmament will also give rise to some 150-200 tonnes of weapons-grade plutonium. In 2000 the USA and Russia agreed to dispose of 34 tonnes each by 2014. While it was initially proposed to immobilize some of the US portion, the general idea is now to fabricate it with uranium oxide as a MOX fuel for burning in existing reactors. A plant is under construction in South Carolina for this fuel fabrication, and meanwhile some trial MOX assemblies (made in France from US military plutonium) are being trialled in a US reactor.

However, Europe has a well-developed MOX capacity and Japan is developing its use. This suggests that weapons-grade plutonium could be disposed of relatively quickly. Input plutonium would need to be

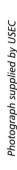

Cylinders containing warhead-derived fuel from Russia

about half reactor-grade and half weapons-grade, but using such MOX as 30% of the fuel in one third of the world's reactor capacity would remove about 15 tonnes of warhead plutonium per year. This would amount to burning 3000 warheads per year to produce 110 billion kWh of electricity.

Over 35 reactors in Europe are licensed to use mixed oxide fuel, and 20 French reactors are using it or licensed to use it as 30% of their fuel. New reactors may be able to run with full MOX cores.

Russia intends to use all of its plutonium as a fuel, burning it in both late-model conventional reactors and particularly in fast neutron reactors. If all the plutonium were used in fast neutron reactors in conjunction with the depleted uranium from enrichment plant stockpiles[2], there would be enough to run the world's commercial nuclear electricity programmes for several decades without any further uranium mining. In Russia a thorium-uranium fuel is being developed which is intended to use weapons-grade plutonium in conventional reactors.

See also: section 5.2 and WNA information paper on Military Warheads as a Source of Fuel.

3.5 THORIUM AS A NUCLEAR FUEL

Most of this book is concerned with uranium as a fuel for nuclear reactors. However, in future, thorium is also likely to be utilized as a fuel for particular reactors. The thorium fuel cycle has some attractive features, and is described further in section 4.7.

Existing neutron efficient reactor designs, such as the Canadian Deuterium Uranium (CANDU) reactor , are capable of operating on a thorium fuel cycle, once they are started using a fissile material such as U-235 or Pu-239. Then the thorium (Th-232) captures a neutron in the reactor to become fissile uranium (U-233), which continues the reaction. However, there are some practical problems with using thorium in this way.

Thorium is about three times as abundant in the Earth's crust as uranium. Australia and India have considerable quantities of thorium, and India is developing its whole nuclear energy programme to make use of it.

See also: WNA information paper on Thorium.

---

[2] When uranium is enriched for a conventional reactor, about seven times more depleted uranium is produced than the enriched product. If uranium is enriched to 93% U-235 for a weapons programme, about 200 times more depleted uranium than enriched product is produced. All this, some 1.2 million tonnes, comprising a very large proportion of all uranium ever mined, is "fertile" material and thus potential fast breeder fuel.

3.6 ACCELERATOR-DRIVEN SYSTEMS

The essence of a conventional nuclear reactor is the controlled fission chain reaction of U-235 and Pu-239. This depends on having a surplus of neutrons to keep it going (a U-235 fission requires one neutron input and produces on average 2.43 neutrons). However, without such a surplus, a nuclear reaction can be sustained by input of neutrons produced by spallation from heavy element targets bombarded by protons in a high-energy accelerator.

If the spallation target is surrounded by a blanket assembly of nuclear fuel, such as fissile isotopes of uranium or plutonium (or thorium which can breed to U-233), there is a possibility of sustaining a fission reaction. This is described as an Accelerator-Driven System (ADS).

In such a subcritical nuclear reactor the neutrons produced by spallation would be used to cause fission in the fuel, assisted by further neutrons arising from that fission, but there are insufficient of the latter to sustain a chain reaction. One then has a nuclear reactor which could be turned off simply by stopping the proton beam, rather than needing to insert control rods to absorb neutrons and make the fuel assembly subcritical. The fuel may be mixed with long-lived fission products or even transuranic nuclides from conventional reactors.

Thus the other role of a subcritical nuclear

Photograph supplied by ISU Photographic Services

Short pulse linear accelerator at Idaho State University, USA.

reactor or ADS is the destruction of heavy isotopes. In the case of atoms of odd-numbered isotopes heavier than thorium-232, they have a high probability of absorbing a neutron and subsequently undergoing nuclear fission, thereby producing some energy and contributing to the multiplication process. Even-numbered isotopes can capture a neutron, perhaps undergo beta decay, and then fission. This process of converting fertile isotopes to fissile ones is called breeding.

Therefore in principle, the subcritical nuclear reactor may be able to convert some long-lived transuranic elements into (generally) short-lived fission products and yield some energy in the process. Here the main benefit would be in making the management and eventual disposal of high-level wastes from nuclear reactors easier and less expensive. Much of the current interest in ADS is in its potential to burn weapons-grade plutonium, as an alternative to using it as mixed oxide fuel in conventional reactors.

3.7 PHYSICS OF A NUCLEAR REACTOR

Neutrons in motion are the starting point for everything that happens in a nuclear reactor.

When a neutron passes near to a heavy nucleus, for example uranium-235, the neutron may be captured by the nucleus and this may or may not be followed by fission. Capture involves the addition of the neutron to the uranium nucleus to form a new compound nucleus. A simple example is U-238 + n ⇨ U-239, which represents formation of the nucleus U-239. The new nucleus may decay into a different nuclide. In this example, U-239 becomes Np-239 after emission of a beta particle (electron), and fissile Pu-239 after emission of another one. But in certain cases the initial capture is rapidly followed by the fission of the new nucleus. Whether fission takes place, and indeed whether capture occurs at all, depends on the velocity of the passing neutron and on the particular heavy nucleus involved.

Nuclear Fission

Fission may take place in any of the heavy nuclei after capture of a neutron. However, low-energy (slow, or thermal) neutrons are able to cause fission only in those isotopes of uranium and plutonium whose nuclei contain odd numbers of neutrons (e.g. U-233, U-235, and Pu-239). Thermal fission may also occur in some other transuranic elements whose nuclei contain odd numbers of neutrons. For nuclei containing an even number of neutrons, fission can only occur if the incident neutrons have energy above about 1 million electron volts (MeV).

The probability that fission or any another neutron-induced reaction will occur is described by the cross-section for that reaction. The cross-section may be imagined as an area surrounding the target nucleus and within which the incoming neutron must pass if the reaction is to take place. The fission and other cross-sections increase greatly as the neutron velocity decreases from around 20,000 km/s to 2 km/s, making the likelihood of some interaction greater. In nuclei with an odd number of neutrons, such as U-235, the fission cross-section becomes very large at the thermal energies of slow neutrons (see Figure 10).

A neutron is said to have thermal energy when it has slowed down to be in thermal equilibrium

Figure 10: Neutron cross-sections for fission of uranium and plutonium

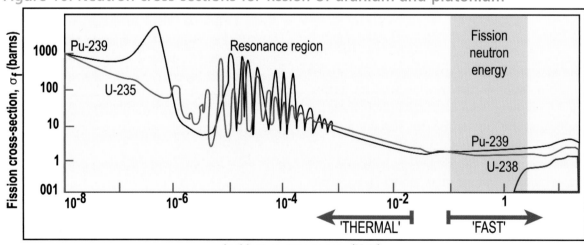

Sources: OECD/NEA 1989, Plutonium fuel – an assessment.
          Taube 1974, Plutonium – a general survey.
1 barn = 10^{-28} m^2, 1MeV = 1.6 x 10^{-13}J

Note that both scales are logarithmic.

Figure 11: Distribution of fission products from uranium-235

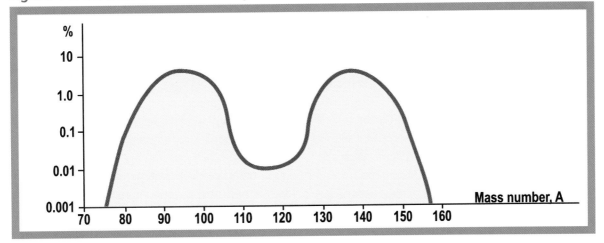

Nuclear Fission – the Process

Using U-235 in a thermal reactor as an example, when a neutron[3] is captured the total energy is distributed amongst the 236 nucleons (protons and neutrons) now present in the compound nucleus. This nucleus is relatively unstable, and it is likely to break into two fragments of around half the mass. These fragments are nuclei found around the middle of the periodic table and the probabilistic nature of the break-up leads to several hundred possible combinations (see Figure 11). Creation of the fission fragments is followed almost instantaneously by emission of a number of neutrons (typically 2 or 3, average 2.5), which enable the chain reaction to be sustained.

About 85% of the energy released is initially the kinetic energy of the fission fragments. However, in solid fuel they can only travel a microscopic distance, so their energy becomes converted into heat. The balance of the energy comes from gamma rays emitted during or immediately following the fission process and from the kinetic energy of the neutrons. Some of the latter are immediate (called prompt neutrons), but a small proportion (0.7% for U-235, 0.2% for Pu-239) is delayed, as these are associated with the radioactive decay of certain

with the surroundings (when the kinetic energy of the neutrons is similar to that possessed by the surrounding atoms due to their random thermal motion). Hence the main application of uranium fission is in thermal reactors fuelled by U-235 and incorporating a moderator such as water to slow the neutrons down. The most common examples of this are Light Water Reactors (Pressurized Water Reactors and Boiling Water Reactors).

As implied previously, high-energy (> 0.1 MeV) neutrons are travelling too quickly to have much interaction with the nuclei in the fuel. We therefore say that the fission cross-section of those nuclei is much reduced at high neutron energies relative to its value at thermal energies (for slow neutrons). However, among the interactions that do occur, fission predominates over capture, so it is possible to use fast neutrons in a fast reactor whose design minimizes the moderation of the high-energy neutrons produced in the fission process. Such a reactor therefore needs highly-enriched fissile material at the start, but it makes very efficient use of it and can breed more fuel from U-238 (see below).

---

3 The chain reaction is started by inserting some beryllium mixed with polonium, radium or other alpha-emitter. Alpha particles from the decay cause the release of neutrons from the beryllium as it turns to carbon-12.

fission products. The longest delayed neutron group has a half-life of about 56 seconds.

The delayed neutron release is the crucial factor enabling a chain reacting system (or reactor) to be controllable and to be able to be held precisely critical. At criticality the chain reacting system is exactly in balance, such that the number of neutrons produced in fissions remains constant. This number of neutrons may be completely accounted for by the sum of those causing further fissions, those otherwise absorbed, and those leaking out of the system. Under these circumstances the power generated by the system remains constant. To raise or lower the power, the balance must be changed (using the control system) so that the number of neutrons present (and hence the rate of power generation) is either reduced or increased. The control system is used to restore the balance when the desired new power level is attained.

The number of neutrons and the specific fission products from any fission event are governed by statistical probability, in that the precise break-up of a single nucleus cannot be predicted. However, conservation laws require the total number of nucleons and the total energy to be conserved. The fission reaction in U-235 produces fission products such as Ba, Kr, Sr, Cs, I and Xe, with atomic masses distributed around 95 and 135. Examples may be given of typical reaction products, such as:

$$U\text{-}235 + n \Rightarrow Ba\text{-}144 + Kr\text{-}90 + 2n + energy$$
$$U\text{-}235 + n \Rightarrow Ba\text{-}141 + Kr\text{-}92 + 3n + 170\ MeV$$
$$U\text{-}235 + n \Rightarrow Te\text{-}139 + Zr\text{-}94 + 3n + 197\ MeV$$

In such an equation the atomic number is conserved (e.g. $235 + 1 = 141 + 92 + 3$), but a small loss in atomic mass may be shown to be equivalent to the energy released. Both the barium and krypton isotopes subsequently decay and form more stable isotopes of neodymium and yttrium, with the emission of several electrons from the nucleus (beta decays). It is the beta decays, with some associated gamma rays, which make the fission products highly radioactive. This radioactivity (by definition!) decreases with time.

The total binding energy released in fission varies with the precise break up, but averages about 200 MeV[4] for U-235, or 3.2×10^{-11} joule. That from Pu-239 is about 210 MeV[4] per fission. (This contrasts with 4 eV or 6.5×10^{-19} J per molecule of carbon dioxide released in the combustion of carbon in fossil fuels.)

About 6% of the heat generated in the reactor core originates from radioactive decay of fission products and transuranic elements formed by neutron capture, mostly the former. This must be allowed for when the reactor is shut down, since heat generation continues after fission stops. It is this decay which makes spent fuel initially generate heat and hence need cooling. Even after one year, typical spent fuel generates about 10 kW of decay heat per tonne, decreasing to about 1 kW/t after ten years.

Neutron Capture: Transuranic Elements and Activation Products

Neutrons may be captured by non-fissile nuclei, and some energy is produced by this mechanism in the form of gamma rays as the compound nucleus de-excites. The resultant new nucleus may become more stable by emitting alpha or beta particles. Neutron capture by one of the uranium isotopes will form what are called transuranic elements, actinides beyond uranium in the periodic table.

Since U-238 is the major proportion of the fuel element material in a thermal reactor, capture

---

[4] These are total available energy release figures, consisting of kinetic energy values (E_k) of the fission fragments plus neutron, gamma and delayed energy releases, which add about 30 MeV.

of neutrons by U-238 and the creation of U-239 is an important process.

- U-239 quickly emits a beta particle to become neptunium-239.
- Np-239 in turn emits a beta particle to become plutonium-239, which is relatively stable.
- Some Pu-239 nuclei may capture a neutron to become Pu-240, which is less stable.
- By further neutron capture, Pu-240 nuclei may in turn form Pu-241.
- Pu-241 also undergoes beta decay to americium-241 (the heart of household smoke detectors).

As already noted, Pu-239 is fissile in the same way as U-235, i.e. with thermal neutrons. It is the other main source of energy in any nuclear reactor and typically contributes about one third of the energy output. The masses of its fission products are distributed around 100 and 135 atomic mass units.

The main transuranic constituents of spent fuel are isotopes of plutonium, neptunium and americium. These are alpha-emitters and have long half-lives, decaying on a similar timescale to the uranium isotopes. They are the reason that spent fuel needs secure disposal beyond the few thousand years or so which might be necessary for the decay of fission products alone.

Apart from transuranic elements in the reactor fuel, activation products are formed wherever neutrons impact on any other material surrounding the fuel. Activation products in a reactor (and particularly its steel components exposed to neutrons) range from tritium (H-3) and carbon-14 to cobalt-60, iron-55 and nickel-63. The latter four radioisotopes create difficulties during eventual demolition of the reactor and affect the extent to which materials can be recycled.

Fast Neutron Reactors

In an idealized Fast Neutron Reactor the fuel in the core is Pu-239 and the abundant neutrons designed to leak from the core would breed more Pu-239 in the fertile blanket of U-238 around the core. A minor fraction of U-238 might be subject to fission, but most of the neutrons reaching the U-238 blanket will have lost some of their original energy and are therefore subject only to capture and the eventual generation of Pu-239. Cooling of the fast reactor core requires a heat transfer medium which has minimal moderation of the neutrons, and hence liquid metals are used, typically sodium or a mixture of sodium and potassium.

Such reactors are up to one hundred times more efficient at converting fertile material than ordinary thermal reactors because of the arrangement of fissile and fertile materials, and there is some advantage from the fact that Pu-239 yields more neutrons per fission than U-235. Although both yield more neutrons per fission when split by fast rather than slow neutrons, this is incidental since the fission cross-sections are much smaller at high neutron energies. Fast neutron reactors may be designed as breeders to yield more fissile material than they consume or to be plutonium burners to dispose of excess plutonium. A plutonium burner would be designed without a breeding blanket, simply with a core optimized for plutonium fuel.

Control of Fission

Fission of U-235 nuclei typically releases 2 or 3 neutrons, with an average of about 2.5. One of these neutrons is needed to sustain the chain reaction at a steady level of controlled criticality; on average the other 1.5 leak from the core region or are absorbed in non-fission reactions. Neutron-absorbing control rods are used to adjust the power output of a reactor. These typically use boron and/or cadmium (both are strong neutron absorbers) and are inserted among the fuel assemblies. When they are slightly withdrawn from their position at criticality, the number of neutrons available for ongoing fission exceeds unity (i.e. criticality is exceeded) and the power level increases. When the power reaches the desired level, the control rods are returned to the critical position and the power stabilizes.

The ability to control the chain reaction is entirely due to the presence of the small proportion of delayed neutrons arising from fission. Without these, any change in the critical balance of the chain reaction would lead to a virtually instantaneous and uncontrollable rise or fall in the neutron population. It is also relevant to note that safe design and operation of a reactor sets very strict limits on the extent to which departures from criticality are permitted. These limits are built in to the overall design.

Neutrons released in fission are initially fast (velocity about 10^9 cm/sec, or energy above 1 MeV), but fission in U-235 is most readily caused by slow neutrons (velocity about 105 cm/sec, or energy about 0.02 eV). A moderator material comprising light atoms thus surrounds the fuel rods in a reactor. Without absorbing too many, it must slow down the neutrons in elastic collisions (compare it with collisions between billiard balls on an atomic scale). In a reactor using natural (unenriched) uranium the only suitable moderators are graphite and heavy water (these have low levels of unwanted neutron absorption). With enriched uranium (i.e. increased concentration of U-235 – see information box in section 4.2), ordinary (light) water may be used as moderator. Water is also commonly used as a coolant, to remove the heat and generate steam.

Other features may be used in different reactor types to control the chain reaction. For instance, a small amount of boron may be added to the cooling water and its concentration reduced progressively as other neutron absorbers build up in the fuel elements. For emergency situations, provision may be made for rapidly adding an excessive quantity of boron to the water.

Commercial power reactors are usually designed to have negative temperature and void coefficients. The significance of this is that if the temperature should rise beyond its normal operating level, or if boiling should occur beyond an acceptable level, the balance of the chain reaction is affected so as to reduce the rate of fission and hence reduce the temperature. One mechanism involved is the Doppler effect, whereby U-238 absorbs more neutrons as the temperature rises, thereby pushing the neutron balance towards subcritical. Another mechanism, in light water reactors, is that the formation of steam within the water moderator will reduce its density and hence its moderating effect, and this again will tilt the neutron balance towards subcritical.

While fuel is in use in the reactor, it is gradually accumulating fission products and transuranic elements, which cause additional neutron absorption. The control system has to be adjusted to compensate for the increased absorption. When the fuel has been in the reactor for three years or so, this build-up in absorption, along with the metallurgical changes induced by the constant neutron bombardment of the fuel materials, dictates that the fuel should be replaced. This effectively limits the burnup to about half of the fissile material, and the fuel assemblies must then be removed and replaced with fresh fuel.

Extending Fission

In naval reactors used for propulsion, where fuel changes are inconvenient, the fuel is enriched to higher levels initially and burnable "poisons" – neutron absorbers – are incorporated. Hence as the fission products and transuranic elements accumulate, the poison is depleted and the two effects tend to cancel one another out. As engineers explore ways to maximize the burnup of commercial reactor fuel, burnable poisons, such as gadolinium, are increasingly used along with increasing enrichment towards 5% U-235.

Another major development gathering impetus is towards burning transuranic elements (plutonium and minor actinides), which are extracted from used fuel. If these remain in discarded used fuel or in wastes from reprocessing used fuel, they will increase the heat load in repositories and complicate the task of designing them. Once the best way of reprocessing used fuel to recover them is resolved, the issue is: how best to fission them? The answer appears to be: in fast neutron reactors, since this minimizes neutron capture and maximizes fission, even yielding some energy. But various thermal reactor and accelerator system methods are also being investigated. (A related question is how to transmute the longer-lived fission products such as technetium-99, iodine-129 and caesium-135 into shorter-lived ones. Here, neutron capture is the objective, and a liberal supply of slow neutrons is required.)

4

THE "FRONT END"
OF THE NUCLEAR
FUEL CYCLE

4.1 MINING AND MILLING OF URANIUM ORE

Uranium minerals are always associated with other elements, such as radium and radon, in radioactive decay series (see Appendix 2). Therefore, although uranium itself is barely radioactive, the ore which is mined must be regarded as potentially hazardous, especially if it is high-grade ore. The radiation hazards involved are similar to those in many mineral sands mining and treatment operations. Canada and Australia, between them, supply over half the world's mined uranium.

Many of the world's uranium mines have been open cut and therefore naturally well ventilated. Ore grades at most mines worldwide are less than 0.5% U_3O_8. The Olympic Dam underground mine in Australia, located in the largest known uranium orebody in the world, has ore grade less than 0.1% U_3O_8. Any underground uranium mine must have an effective ventilation system.

Canada's McArthur River and Cigar Lake mines are in very high-grade ore and require special remote-control techniques for mining.

The mined ore (i.e. rock containing economically recoverable concentrations of uranium) is crushed and ground. The resulting slurry is then leached, usually with sulphuric acid, to dissolve the uranium (together with some other metals). The solids remaining after the uranium is extracted are known as tailings. They are pumped as a slurry to the tailings dam, which is engineered to retain them securely. Tailings contain most of the radioactive material in the ore, such as radium.

Some newer mines are "in situ leaching" (ISL) operations, with recovery of the uranium from the sandy ore taking place underground. A slightly acidic and heavily oxygenated solution is circulated through boreholes and the uranium is extracted in plant at the surface, with the liquor being recirculated.

In each case the leach liquor goes through a solvent extraction or ion exchange process followed by precipitation to remove the uranium from solution as a bright yellow precipitate ("yellowcake"). After high-temperature drying, the uranium oxide (U_3O_8), now khaki in colour, is packed into 200-litre drums for shipment. The radiation level one metre from such a drum of freshly processed U_3O_8 is about half that received by a person from cosmic rays on a commercial jet flight.

In Australia all these operations are undertaken under the Australian Code of Practice and Safety Guide: *Radiation Protection and Radioactive Waste Management in Mining and Mineral Processing* (2005), administered by state governments. In Canada the Canadian Nuclear Safety Commission regulations apply. In both countries there are strict health standards for gamma radiation and radon gas exposure[1] as well as for ingestion and inhalation of radioactive materials. In other countries similar regulations are in place. Standards apply to both workers and members of the public.

The gamma radiation comes principally from isotopes of bismuth and lead. The radon gas emanates from the rock (or tailings) as radium decays[2]. It then decays itself to (solid) radon daughters, which are energetic alpha-emitters. Radon occurs in most rocks and traces of it are in the air we all breathe. However, at high concentrations it is a health hazard since its short half-life means that disintegrations giving off alpha particles are occurring relatively frequently. Alpha particles discharged in the lung can later give rise to lung cancer.

---

[1] 20 mSv/yr averaged over 5 years is the maximum allowable radiation dose rate for workers, including radon (and radon daughters) dose. This is in addition to natural background and excludes medical exposure. See also Appendix 1 and glossary for definitions.
[2] "Radon" here normally refers to Rn-222. Another isotope, Rn-220 (known as "thoron"), is given off by thorium, which is a constituent of many Australian mineral sands. See also Appendix 2.

A number of precautions are taken at any mine to protect the health of workers, and those at a uranium mine are slightly greater:

• Dust is controlled to minimize inhalation of gamma or alpha-emitting minerals. In practice, dust is the main source of radiation exposure in a uranium mine. At Ranger it typically contributes about 2 mSv/yr to a worker's annual dose (see also Table 13).

• Radiation exposure of workers in the mine, plant and tailings areas is limited. In practice, direct radiation levels from the ore and tailings are usually so low that it would be difficult for a worker to come anywhere near the allowable annual dose. However, in some Canadian mines the dose would be so high that people are excluded from the workings.

• Radon daughter exposure is limited in an open-cut mine because there is sufficient natural ventilation and the radon level seldom exceeds 1% of the levels allowable for continuous occupational exposure. In an underground mine a good forced ventilation system is required to achieve the same result – at Olympic Dam (Australia) radiation doses in the mine are kept very low, with an average of less than 1 mSv/yr. In Canada, doses average about 3 mSv/yr.

Strict hygiene standards, similar to those in a lead smelter, are imposed on workers handling the uranium oxide concentrate. If it is ingested it has a chemical toxicity similar to that of lead oxide[3]. Respiratory protection is used in particular areas identified by air monitoring or where there could be a hazard.

These precautions with respect to radon are a relatively new phenomenon. From the 15th century, many miners who had worked underground in the mountains near the present border between Germany and the Czech Republic contracted a mysterious illness, and many died prematurely. In the late 1800s the illness was diagnosed as lung cancer, but it was not until 1921 that radon gas was suggested as the possible cause. Although this was confirmed by 1939, between 1946 and 1959 a lot of underground uranium mining took place in the USA without the precautions which might have become established as a result of the European experience. In the early 1960s a higher than expected incidence of lung cancer began to show up among miners who smoked. The cause was then recognized as the emission of alpha particles from radon and, more importantly, its solid daughter products of radioactive decay. The miners concerned had been exposed to high levels of radon 10-15 years earlier, accumulating radiation doses well in excess of present recommended levels.

The small, unventilated uranium "gouging" operations in the USA which led to the greatest health risk are a thing of the past. In the last 40 years, individual mining operations have been larger, and efficient ventilation and other precautions now protect underground miners from these hazards. Open cut mining of uranium virtually eliminates the danger. There has been no known case of illness caused by radiation among uranium miners in Australia or Canada. While this may be partly due to the lack of detailed information on occupational health from operations in the 1950s, it is clear that no major occupational health effects have been experienced in either country.

After mining is complete most of the orebody, with virtually all of the radioactive radium, thorium and actinium materials, will end up in the tailings dam[4]. Hence radiation levels and

---

[3] Both lead and uranium are toxic and affect the kidney. The body progressively eliminates most Pb or U via urine.

radon emissions from tailings will probably be significant. In the unlikely event of someone setting up camp on top of the material, they could eventually receive a radiation dose exceeding international standards, just as they could from outcropping orebodies. Therefore, the tailings need to be covered over with enough rock, clay and soil to reduce both gamma radiation levels and radon emanation rates to levels near those naturally occurring in the region. A vegetation cover can then be established.

Radon emanation from tailings during mining and before they are covered is sometimes seen as an environmental hazard to be considered in isolation from the traces of radon which are continually emitted by uranium and thorium minerals present in most rock and soil. In fact, apart from the local hazards mentioned above, any regional increase in radon release due to mining operations is very small and not measurable outside the mine lease (see also notes on radiation in section 7.4).

Process water, from which tailings solids have settled out, contains radium and other metals that would be undesirable in the outside environment. This water is retained and evaporated so that the contained metals are retained in safe storage, as in an orebody. In fact process water is never released to natural waterways, but is stored in tailings retention area, and evaporated from there or treated for re-use.

At Ranger, rainfall run-off is segregated in accordance with water quality, and high quality water from relatively undisturbed catchments is released during flood times[5]. Poorer quality water is retained on site and treated.

4.2 THE NUCLEAR FUEL CYCLE

Fuel cycles describe the way in which fuel gets to where it is used to provide energy and what happens to it afterwards. The "front end" of the nuclear fuel cycle covers all the stages from uranium mining to burning of the fuel in the reactor. The "back end" refers to all stages subsequent to removal of used fuel from the reactor.

All aspects of obtaining and preparing the fuel, using it, and managing used fuel together make up what is known as the nuclear fuel cycle. As the term suggests it was originally the intention with nuclear power to recycle the unused part of the spent fuel so that it is incorporated into the fresh fuel elements. However, more commonly this is not done today because fresh supplies of uranium are relatively inexpensive.

Unlike coal, uranium as mined cannot be fed directly into a power station. It has to be purified, isotopically concentrated (usually) and made up into special fuel rods. Figure 12 shows the so-called "open fuel cycle" for nuclear power, which is the system as it stands today in most countries using the most common kinds of reactors.

Starting in **uranium mines**, ore is mined and milled to produce uranium in the form of uranium oxide concentrate, commonly known as U_3O_8. This material, a khaki-coloured powder after drying, is sold to customers and shipped from the mine. It has the same isotopic ratio as the ore, where U-235 is present to the extent of about 0.7%. Apart from traces of U-234, the rest is a heavier isotope of uranium – U-238. Most reactors, including the common light water type (LWR), cannot run on this natural uranium, so the proportion of U-235

---

[4] About 95% of the radioactivity in the ore is from the U-238 decay series (see Appendix 2), totalling about 450 kBq/kg in ore with 0.3% U_3O_8. The U-238 series has 14 radioactive isotopes in secular equilibrium, thus each represents about 32 kBq/kg (irrespective of the mass proportion). When the ore is processed, the U-238 and the very much smaller masses of U-234 (and U-235) are removed. The balance becomes tailings, and at this point has about 85% of its original intrinsic radioactivity. However, with the removal of most U-238, the two short-lived decay products, Th-234 & Pa-234, soon disappear, leaving the tailings with a little over 70% of the radioactivity of the original ore after several months. The controlling long-lived isotope then becomes Th-230, which decays with a half-life of 77,000 years to radium-226, followed by radon-222 (Alex Zapantis, Supervising Scientist Group, Australia).
[5] Radionuclide levels are not to exceed drinking water standards.

Figure 12: The open fuel cycle

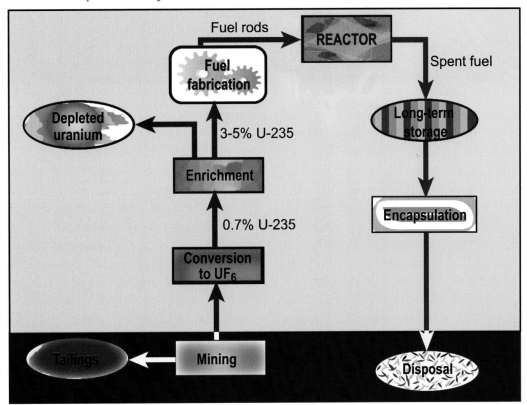

must be increased to between 3% and 5%. This process is called **enrichment**.

Enrichment is a fairly high-technology physical process which requires the uranium to be in the form of a gas. The simplest way to achieve this is to convert the uranium oxide to uranium hexafluoride, which is a gas at little more than room temperature (actually at 56°C). This form of uranium is commonly referred to as UF$_6$ or "hex". Hence the first destination of uranium oxide concentrate from a mine is a **conversion plant** where it is purified and converted to uranium hexafluoride.

The UF$_6$ is then fed to an **enrichment plant**[6], which increases the proportion of the fissile U-235 isotope about five- or six-fold from the 0.7% of U-235 found in natural uranium. In this

physical process about 85% of the natural uranium feed is rejected as "depleted uranium", or "tails" (mainly U-238), which is stockpiled[7]. Thus, after enrichment about 15% of the original quantity is available as enriched uranium containing about 3.5% or more U-235.

The enrichment methods now in use are based on the slight difference in atomic mass of U-235 and U-238. Some 40% of the installed capacity relies on the gaseous diffusion process, where the UF$_6$ gas is passed through a long series of membrane barriers which allow the lighter molecules with U-235 through faster than the U-238 ones. Other more modern plants use high-speed centrifuges to separate the molecules of the two isotopes.

---

[6] Enrichment was originally undertaken using the expensive and energy intensive gaseous diffusion process. Newer plants are mostly based on very much more efficient gas centrifuge technology. The next generation of enrichment plants may use advanced laser technology.
[7] This material cannot be used in current types of reactors and its only significant use is as a feed for fast breeder reactors, or to dilute ex-military uranium (see sections 3.5. & 4.4). It is stored as UF$_6$ in steel cylinders as a liquid or solid. Usually less than 0.3% U-235 remains in it.

Uranium Enrichment

The two main enrichment (or isotope separation) processes are diffusion (gas diffusing under pressure through a membrane containing microscopic pores) and centrifugation. In each case a very small amount of isotope separation takes place in one pass through the process. Hence repeated separations are undertaken in successive stages, arranged in a cascade. The product from each stage becomes feed for the next stage above, and the depleted material is added to the feed for the next stage below. The stages above the initial feed point thus become the enriching section, and those below are the stripping section; each stage has a double feed (product from below and depleted from above). Ultimately, the enriched product is about one sixth or one seventh the amount of depleted material, so that the product end of the cascade tends to have more stages. The depleted material, drawn off at the bottom of the stripping section, is commonly called tails, and the residual U-235 concentration in the tails is the tails assay.

A fuller description of enrichment processes is in the WNA information paper on Uranium Enrichment.

The separating power of the cascade, or of each stage, is described in terms of flow capacity and enriching ability, using the separative work unit (SWU) to quantify it. This is dimensionally a mass unit, though it indicates energy (for a particular plant, energy consumption may be described in kWh per SWU). Since feed or product quantities are measured in tonnes or kilograms, SWUs are also described similarly.

For instance, to produce one kilogram of uranium enriched to 3.5%, U-235 requires

4.3 SWU if the plant is operated at a tails assay of 0.30%, or 4.8 SWU if the tails assay is 0.25% (thereby requiring only 7 kg instead of almost 8 kg of natural uranium feed). Here, and in the following paragraph, kg SWU units are implied.

About 100,000 to 120,000 SWU is required to enrich the annual fuel loading for a typical 1000 MWe light water reactor. Enrichment costs are related to electrical energy used. The gaseous diffusion process consumes up to 2400 kWh (8600 MJ) per SWU, while gas centrifuge plants require only about 50 kWh/SWU (180 MJ). Despite this, competition between the commercial enrichment plants compels comparable prices to be charged, and the relative capital costs and ages of the enrichment plants may compensate the large difference in the electric power components.

The diffusion process relies on a difference in average velocity of the two types of UF_6 molecules to drive the lighter ones more readily through holes in the membranes. Each stage consists of a compressor, a diffuser and a heat exchanger to remove the heat of compression. The enriched UF_6 product is withdrawn from one end of the cascade, and the depleted UF_6 is removed at the other

"Eurodif" diffusion enrichment plant at Tricastin, France

end. The gas must be processed through some 1400 stages to obtain a product with a concentration of 3% to 4% U-235. Diffusion plants typically have a small amount of separation through one stage (hence the large number of stages), but are capable of handling large volumes of gas.

Centrifuge enrichment relies on the simple mass difference of the molecules coupled with the square of the peripheral velocity in a rapidly rotating cylinder (the centrifuge rotor). Countercurrent movement of gas within the rotor, proportional to its height, enhances this effect. The gas is fed into a series of evacuated cylinders, each containing a rotor about 1 to 2 m long and 15 to 20 cm diameter. When the rotors are spun rapidly, the heavier molecules with U-238 increase in concentration towards the cylinder's outer edge, leaving a corresponding increase in concentration of molecules with U-235 near the centre. The countercurrent flow enables enriched product to be drawn off axially.

To obtain efficient separation of the two isotopes, centrifuges rotate at very high speeds, typically 50,000 to 70,000 rpm, with the outer wall of the spinning cylinder moving at between 400 to 500 m/sec, to give a million times the acceleration of gravity. There are considerable materials and engineering challenges in producing such equipment.

Although the volume capacity of a single centrifuge is much smaller than that of a single diffusion unit, its ability to separate isotopes is much greater. Centrifuge stages normally consist of a large number of centrifuges in parallel. Such stages are then arranged in cascade similarly to those for diffusion. In the centrifuge process, the number of stages may only be 10 to 20, instead of the 1000 or more required for diffusion.

Laser isotope separation processes have been a focus of interest for some time. They promise lower energy inputs, lower capital costs and lower tails assays, hence significant economic advantages. None of these processes is yet ready for commercial use. In addition, the US AVLIS process, into which billions of dollars had been invested and which was thought to be in an advanced stage of development, was cancelled in 1999.

Laser processes utilize the very precise beam frequencies characteristic of lasers. Such frequencies are equivalent to defined energies. The interaction of the laser beam with gas or vapour enables it to exploit the excitation or ionization of isotope-specific atoms in the vapour. It may then be possible to separate molecules containing a desired isotope by utilizing a second physical process applicable only to the excited or ionized molecule. For instance a tuned laser of very specific energy might convert UF_6 molecules containing U-235 atoms to solid UF_5, by breaking the molecular bond holding the sixth fluorine atom. This then enables the UF_5 to be separated from the unaffected UF_6 molecules containing U-238 atoms, hence achieving a separation of isotopes.

Laser separation processes may use either atomic or molecular gases or vapours. AVLIS is an atomic process. A molecular laser separation process, SILEX, utilizing uranium in the form of UF_6, is currently under development in Australia.

Enriched uranium then goes on to a fuel fabrication plant, where the reactor fuel elements are made. The UF_6 is converted to UO_2, which is formed into small cylindrical pellets about 2 cm long and 1.5 cm in diameter. These are heated to high temperature to form hard ceramic pellets, and then loaded into zirconium alloy or stainless steel tubes about 4 m long to form fuel rods. The rods are assembled into bundles about 30 cm square to form reactor fuel assemblies. Fuel assemblies of this type are used to power light water power reactors, currently the most popular design (see Table 5). A 1000 MWe reactor has about 75 tonnes of fuel in it.

Canadian CANDU reactors have a different design and run on natural (i.e. unenriched) uranium. Instead of a single large pressure vessel containing the core, they have multiple (e.g. 300-600) horizontal pressure tubes, each containing fuel and heavy water coolant. The pressure tubes extend through the reactor vessel, or calandria, which contains the heavy water moderator[8]. CANDU fuel bundles are only 10 cm diameter and 50 cm long.

Inside all kinds of operating **reactors** a fission chain reaction occurs in the fuel rods, as described in sections 3.1 and 3.7. Fast neutrons are slowed by the water, heavy water or graphite moderator so that they can cause fission. Neutron-absorbing control rods are inserted or withdrawn to regulate the speed of the reaction. Heat from the fission reaction is conveyed from the reactor core by the coolant and is used to make steam, which in turn is used to generate electricity.

In a light water reactor the fuel stays in the reactor for about three to four years, generating heat from fission of both the U-235 and the fissile plutonium (e.g. Pu-239), which is formed there from U-238. Progressively over

a few years, the level of fission products and other neutron-absorbers builds up so that they interfere with the fission chain reaction, and the used fuel assemblies are therefore removed. About one third of the fuel may be changed each year or more.

When removed, used fuel is hot and radioactive. It is therefore stored under water to remove the heat and to provide shielding from radiation, pending the next step. This may be reprocessing in the case of such countries as the UK, France and Japan, which have chosen to "close" the fuel cycle, or it may be final disposal in the case of such countries as the USA, Canada and Sweden, which have chosen the "open fuel cycle". Storage is initially at the reactor site. The used fuel may then be transferred elsewhere, or to an engineered dry storage facility.

Earlier generations of reactors, such as are still operating in the UK, use uranium metal fuel instead of uranium oxide and are gas-cooled. For these reactors reprocessing operations have been going on for some time, so that the fuel elements are not held very long in cooling ponds. This, and the corresponding arrangement for light water reactors, is illustrated by the more complex diagram in Figure 13, which is known as the "closed fuel cycle" system.

In the closed fuel cycle for light water reactors, fuel is supplied in exactly the same way as before. Starting with uranium mines and mills the uranium goes through conversion, enrichment, and fuel fabrication to the reactor.

But after being removed from the reactor the used fuel rods are put through a **reprocessing plant**, where they are chopped up and dissolved in acid. Various chemical processes recover and separate the two valuable

---

[8] Heavy water, or deuterium oxide, contains deuterium, which is an isotope of hydrogen having a neutron in the nucleus.

Figure 13: The closed fuel cycle

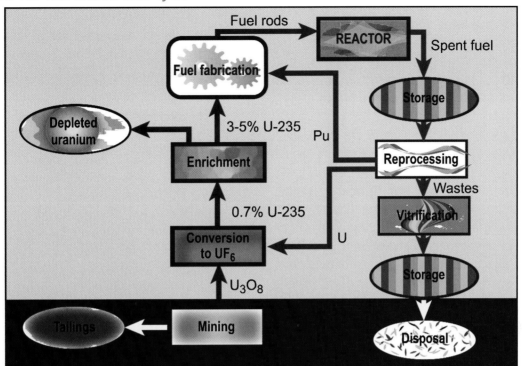

components: plutonium and unused uranium. This leaves about 3% of the fuel as separated high-level waste. After solidification it is reduced to a small volume of highly radioactive material in solid form suitable for permanent disposal (see also sections 5.2 and 5.3).

About 96% of the uranium which goes into the reactors emerges again in the spent fuel, albeit depleted to less than 1% U-235. As shown in Figure 16 some of what has been used up was converted into heat and radioactive fission products and some into plutonium and other actinide elements. Hence reprocessing spent fuel has some economic benefits in recovering the unused uranium and the plutonium which has been generated and not burned in the reactor. It also substantially reduces the volume of material to be disposed of as high-level waste, which has economic benefit.

Plutonium (which is reactor-grade, not weapons-grade) comprises about 1% of the spent fuel. It is a mixture of isotopes and makes a very good nuclear fuel which needs no

enrichment process. It can be mixed with depleted uranium, made into fuel rods in a MOX fuel fabrication plant, and put back into the reactor as fresh fuel (see section 5.2). Alternatively, it could be used to fuel future breeder reactors (see section 4.4 below).

The recovered uranium can go back to be enriched (via conversion), and then introduced as fresh fuel for a reactor. The closed fuel cycle is thus a more efficient system for making maximum use of the uranium dug out of the ground (by about 30%, in energy terms) and that is why the industry originally favoured this approach. However, due largely to many years of low uranium prices (mid 1980s to about 2003) and fears about separating plutonium, plans for widespread reprocessing of spent reactor fuel have not eventuated. France, Germany, UK, Switzerland, Russia and Japan are proceeding with the closed fuel cycle for oxide fuels, and across Europe over 35 reactors are licensed to load 20% to 50% of their core with MOX fuel containing up to 7% reactor-grade plutonium.

4.3 ADVANCED REACTORS

Today's nuclear reactor technology is distinctly better than that represented by most of the world's operating plants, and the first third-generation advanced reactors are now in service in Japan.

Several generations of reactors are commonly distinguished. Generation I reactors were developed in the 1950s and 1960s, and outside the UK none are still running today. Generation II reactors are typified by the present US fleet and most of those reactors in operation elsewhere. Generation III (and III+) are the advanced reactors discussed here. The first few are in operation in Japan and others are under construction or ready to be ordered. Generation IV designs are still on the drawing board and will not be operational before 2020 at the earliest.

About 85% of the world's nuclear electricity is generated by reactors derived from designs originally developed for naval use. These and other second-generation nuclear power units have been found to be safe and reliable, but they are being superseded by better designs.

Reactor suppliers in North America, Japan and Europe have several new nuclear reactor designs either approved or at advanced stages of planning, and other designs at a research and development stage. These incorporate safety improvements and will also be simpler to build, operate, inspect and maintain, thus increasing their overall reliability and economy.

The new generation reactors:

- have a standardized design for each type to expedite licensing, reduce capital cost and reduce construction time.
- are simpler and more rugged in design, easier to operate and less vulnerable to operational upsets.
- have higher availability and longer operating life – typically 60 years.
- further reduce the possibility of core melt accidents.
- have higher burnup to reduce fuel use and the amount of waste.
- will be economically competitive, some of them in a range of sizes.

The greatest change from most designs now operating is that many new nuclear plants will have more "passive" safety features, which rely on gravity, natural convection and so on, requiring no active controls or operational intervention to avoid accidents in the event of malfunction. They will allow operators more time to remedy problems, which will provide greater assurance regarding containment of radioactivity in all circumstances.

A separate line of development epitomizing passive safety is of small high temperature reactors with refractory fuel capable of withstanding very high temperatures, and cooled by helium. These are put forward as "intrinsically safe", in that no emergency cooling system is needed, and in the event of a problem the units can be left to themselves. Being small, the high surface to volume ratio enables dissipation of heat naturally (see also section 4.4 below).

In the USA, the federal Department of Energy (DOE) and the commercial nuclear industry have developed four advanced reactor types. Two of them fall into the category of large evolutionary designs, which build directly on the experience of operating light water reactors in the USA, Japan and Western Europe. These reactors are in the 1300-megawatt range.

One is an advanced boiling water reactor (ABWR), examples of which are in commercial operation in Japan, with more under construction there and in Taiwan. The other type, System 80+, is an advanced pressurized water reactor (PWR), which was ready for commercialization but is not now being promoted for sale. However, eight System 80 reactors in South Korea incorporate many design features of the System 80+, which is the basis of the Korean Next Generation Reactor programme, specifically the APR-1400. The APR-1400 is expected to be in operation soon after 2010 and marketed worldwide.

The US Nuclear Regulatory Commission (NRC) gave final design certification for both the ABWR and System 80+ in 1997, noting that they exceeded NRC "safety goals by several orders of magnitude". The ABWR has also been certified as meeting the stringent European Utility Requirements (EUR) of French and German utilities for advanced reactors.

Another more innovative US advanced reactor is smaller – 600 MWe – and has passive safety features (its projected core damage frequency is nearly 1000 times less than today's NRC requirements). The Westinghouse AP-600 gained final design certification from the NRC in 1999.

These NRC approvals were the first such generic certifications to be issued and are valid for 15 years. As a result of an exhaustive public process, safety issues within the scope of the certified designs have been fully resolved and hence will not be open to legal challenge during licensing for particular plants. Utilities will be able to obtain a single NRC licence to both construct and operate a reactor before construction begins.

Separate from the NRC process and beyond its immediate requirements, the US nuclear industry has selected one standardized design in each category, the large ABWR and the medium-sized AP-600, for detailed first-of-a-kind engineering (FOAKE) work. The US$ 200

million programme was half-funded by the DOE. It meant that prospective buyers would have firm information on construction costs and schedules.

The Westinghouse AP-1000, scaled-up from the AP-600, received final design certification from the NRC in 2005 – the first Generation III+ reactor to do so. It represents the culmination of a 1300 man-year and $440 million design and testing programme. Capital costs appear competitive and modular design should reduce construction time to 36 months. The 1100 MWe AP-1000 generating costs are expected to be below US$ 3.5 cents/kWh, and it has a 60-year operating life. It is under active consideration for building in China, Europe and USA, and is capable of running on a full MOX core if required.

General Electric has developed the Economic Simplified Boiling Water Reactor (ESBWR) of 1390 MWe with passive safety systems, from its ABWR design. This then grew to 1550 MWe and has been submitted for NRC design certification in the USA. Design approval is expected by 2007. It is favoured for early US construction.

Another international project of US origin which is a few years behind the AP-1000 is the International Reactor Innovative & Secure (IRIS). Westinghouse is leading a wide consortium developing it as an advanced Generation III project. IRIS is a modular 335 MWe pressurized water reactor with integral steam generators and primary coolant system all within the pressure vessel. Fuel is initially similar to present LWRs with 5% enrichment, but is designed ultimately for 10% enrichment or equivalent MOX core and higher burnup with an 8-year cycle. IRIS could be deployed in the next decade, and US design certification is envisaged by 2010.

See also: WNA information paper on Generation IV Reactors.

Table 8: Advanced nuclear reactors being marketed

Country & developer	Reactor	Size - MWe	Design progress	Main features (improved safety in all)
US-Japan (GE - Hitachi - Toshiba)	ABWR	1300	Commercial operation in Japan since 1996-1997. In US: NRC certified 1997, FOAKE.	• Evolutionary design • More efficient, less waste • Simplified construction (48 months) and operation
USA (Westinghouse)	AP-600 AP-1000 (PWR)	600 1100	AP-600: NRC certified 1999, FOAKE. AP-1000 - NRC certification 2005.	• Simplified construction and operation • 3 years to build • 60-year plant life
France-Germany (Framatome ANP)	EPR (PWR)	1600	Future French standard; French design approval. Being built in Finland. US version being developed.	• Evolutionary design • High fuel efficiency • Low cost electricity
USA (GE)	ESBWR	1550	Developed from ABWR. Under certification in USA.	• Evolutionary design • Short construction
Japan (utilities + Westinghouse, Mitsubishi)	APWR	1500	Basic design in progress. Planned for Tsuruga.	• Hybrid safety features • Simplified construction and operation
South Korea (derived from Westinghouse)	APR 1400 (PWR)	1450	Design certification 2003. First units expected to be operating about 2012.	• Evolutionary design • Increased reliability • Simplified construction (48 months) and operation
Germany (Framatome ANP)	SWR-1000 (BWR)	1200	Under development; pre-certification in USA.	• Innovative design • High fuel efficiency
Russia (Gidropress)	V-448 (PWR)	1500	Replacement for Leningrad and Kursk plants.	• High fuel efficiency • Enhanced safety
Russia (Gidropress)	V-392 (PWR)	950	Two being built in India; likely bid for China.	• Evolutionary design • 60-year plant life
Canada (AECL)	ACR	700 1000	ACR-1000 proposed for UK; undergoing certification in Canada.	• Evolutionary design • Low-enriched fuel • Light water cooling
HTRs South Africa (Eskom, Westinghouse)	PBMR	165 (module)	Demo plant due to start constructing 2006.	• Modular plant, low cost • Direct cycle gas turbine • High fuel efficiency
USA-Russia (General Atomics - OKBM)	GT-MHR	285 (module)	Under development in Russia by multinational joint venture.	• Modular plant, low cost • Direct cycle gas turbine • High fuel efficiency

In Japan, the first two ABWRs have been operating since 1996 and are expected to have a 60-year life. Two more started up in 2004 and 2005. Several more are under construction in Japan and also Taiwan.

Photograph supplied by GE Electric

Representation of the ABWR

A large (1500 MWe) Advanced PWR (APWR) is being developed by four utilities together with Westinghouse and Mitsubishi. It is intended as the basis for the next generation of Japanese PWRs. In addition, Mitsubishi is participating in development of Westinghouse's AP-1000 reactor.

In South Korea, the APR-1400 Advanced PWR design has evolved from the US System 80+ and has been known as the Korean Next-Generation Reactor. Design certification by the Korean Institute of Nuclear Safety was awarded in 2003. The first of these 1450 MWe reactors is expected to be operating about 2012.

In Europe, new designs under development meet European requirements (EUR) . The first two are ready for commercial deployment.

Framatome ANP has developed a large (1600 MWe and up to 1750 MWe) European Pressurized water Reactor (EPR), which received French design approval in 2004. It is derived from the French N4 and German Konvoi types. It has the highest thermal efficiency of any light water reactor and a 60-year life. The first unit is being built in Finland and the second is scheduled for France. A US version of the EPR is also undergoing review in the USA, with the intention of a design certification application in 2007.

Together with German utilities and safety authorities, Framatome ANP is also developing another evolutionary design, the SWR 1000, a 1250 MWe BWR. The design was completed in 1999 and development continues, with US design certification being sought.

In Russia, several advanced reactor designs have been developed, and all are advanced PWRs with passive safety features.

Gidropress 1000 MWe V-392 (advanced VVER-1000) units with enhanced safety are planned for Novovoronezh and are being built in India. A transitional VVER-91 was developed with western control systems, and two are being built in China.

The VVER-1500 V-448 model is being developed by Gidropress, and two units each are planned as replacement plants for Leningrad and Kursk. It will have high burnup and enhanced safety. Design is expected to be complete in 2007 and the first units commissioned in 2012-2013.

Small floating nuclear power plants are also being developed.

Canada has designs under development which are based on its reliable CANDU-6 reactors; the most recent of these reactors are operating in China.

The main one is the Advanced CANDU Reactor (ACR). While retaining the low-pressure heavy water moderator, it incorporates some features of the pressurized water reactor, including low-enriched fuel. Adopting light water cooling and a more compact core reduces capital cost, and because the reactor is run at higher temperature and coolant pressure, it has higher thermal efficiency.

The ACR-700 is 750 MWe but is physically much smaller, simpler, more efficient as well as 40% cheaper than the CANDU-6. The ACR-1000 of 1200 MWe offers economies of scale and is now the focus of attention. The ACR will run on low-enriched uranium (about 1.5%-2.0% U-235) with high burnup, extending the fuel life by about three times and reducing high-level waste volumes accordingly. Units will be assembled from prefabricated modules, allowing construction time to be cut to 3.5 years. ACR units can be built singly but are optimal in pairs.

ACR is moving towards design certification in Canada, with a view to following in China, the USA and the UK.

India is developing the Advanced Heavy Water reactor (AHWR) as the third stage in its plan to utilize thorium to fuel its overall nuclear power programme. The AHWR is a 300 MWe reactor moderated by heavy water at low pressure. It is designed to be self-sustaining in relation to U-233 bred from Th-232 (see section 4.5).

See also: WNA information papers on Advanced Nuclear Power Reactors and Small Nuclear Power Reactors.

4.4 HIGH TEMPERATURE REACTORS

Building on the experience of several innovative reactors built in the 1960s and 1970s, new high-temperature gas-cooled reactors (HTRs) are being developed which will be capable of delivering high-temperature helium (up to 950°C) either for industrial application or directly to drive gas turbines for electricity (the Brayton cycle) with about 48% thermal efficiency possible. Technology developed in the last decade makes HTRs more practical than in the past, though the direct cycle means that there must be high integrity of fuel and reactor components.

Fuel for these reactors is in the form of particles less than a millimetre in diameter. Each has a kernel of uranium oxycarbide, with the uranium enriched up to 17% U-235. This is surrounded by layers of carbon and silicon carbide, giving a containment for fission products which is stable to 1600°C or more.

There are two ways in which these particles are arranged: in blocks or hexagonal "prisms" of graphite; or in billiard ball-sized "pebbles" of graphite encased in silicon carbide, each with about 15,000 fuel particles and 9 g uranium. Both have a high level of inherent safety, including a strong negative temperature coefficient (whereby fission slows as temperature rises).

South Africa's Pebble Bed Modular Reactor (PBMR) is being developed by a consortium led by the utility Eskom but involving Westinghouse, and drawing on German expertise. It aims for a step change in safety and economics. Modules with a direct-cycle gas turbine generator will be of 165 MWe and have a thermal efficiency of about 42%. Up to 450,000 fuel pebbles recycle through the graphite-lined reactor continuously (about six times each) until they are expended, giving an average enrichment in the fuel load of 4% to 5%. Each unit will finally discharge about 19 tonnes/yr of spent pebbles to ventilated on-site storage bins. A similar unit has been announced for China, and the two countries have forged a technical cooperation agreement.

China has had a small pebble bed reactor running successfully since 2000 at Tsinghua University INET. Construction of a larger version, the 200 MWe HTR-PM, was approved in 2005. This will use 9% enriched fuel (520,000 elements) in an annular core. It will drive a steam cycle turbine. This demonstration reactor at Weihei in Shandong province is to pave the way for an 18-module full-scale power plant on the same site, also using the steam cycle. Plant life is envisaged as 60 years with 85% load factor. Huaneng, one of China's major generators, is the lead organization involved in the demonstration unit with 50% share. Start-up is scheduled for 2010. The HTR-PM rationale is both to eventually replace conventional reactor technology for power and to provide for future hydrogen production.

A larger US design, the Gas Turbine – Modular Helium Reactor (GT-MHR), will be built as modules of 285 MWe, each directly driving a gas turbine at 48% thermal efficiency. It is being developed by General Atomics in partnership with Russia's Minatom, supported by Framatome ANP and Fuji (Japan). Initially it will be used to burn pure ex-weapons plutonium at Tomsk in Russia.

HTRs can potentially use thorium-based fuels, such as high-enriched uranium with Th, U-233 with Th, and Pu with Th. Most of the experience with thorium fuels has been in HTRs.

4.5 FAST NEUTRON REACTORS

Fast neutron reactors are a different technology from those considered so far. They generate power from plutonium by much more fully utilizing the uranium-238 in the reactor fuel assembly, instead of needing just the fissile U-235 isotope used in most reactors (see also section 3.7). If they are designed to produce more plutonium than they consume, they are called Fast Breeder Reactors (FBR). If they are net consumers of plutonium they are sometimes called "burners". For many years the focus was on the potential of this kind of reactor to produce more fuel than they consume, but with low uranium prices (mid 1980s to about 2003) and the need to dispose of plutonium from military weapons stockpiles, the main short-term interest is in their role as incinerators.

Several countries have research and development programmes for FBRs, which are, generically, fast neutron reactors. Over 300 reactor-years of operating experience has been gained on this type of plant. However, the programmes have faltered, and significant technical and materials problems were encountered. The French programme was derailed by political decision, the Japanese programme was suspended due to a coolant leak, and only the Russian programme continues with any vigour (see Table 9).

In the closed fuel cycle (Figure 13) it can be seen that conventional reactors give rise to three "surplus" materials: depleted uranium (from enrichment), plutonium (from neutron capture in the reactor core, separated in reprocessing), and reprocessed uranium (with around 1% U-235). The fast neutron reactor has no moderator and uses plutonium as its basic fuel since it fissions sufficiently with fast neutrons to keep going. At the same time the number of neutrons produced per fission is 25% more than from uranium, and this means

Figure 14: The fast breeder fuel cycle

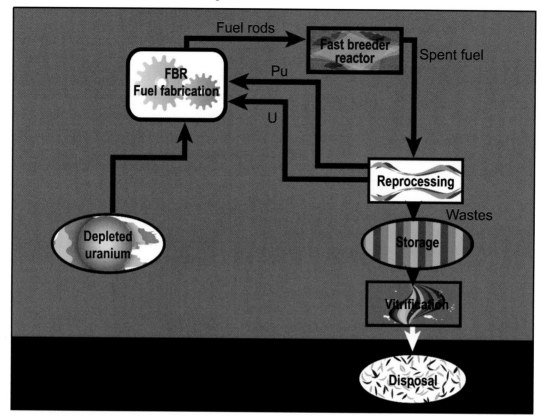

that there are enough (after losses) not only to maintain the chain reaction but also to convert some depleted uranium – basically U-238 comprising a "fertile blanket" around the core – into fissile plutonium. In other words the fast reactor "burns" and can "breed" plutonium[9], as shown in Figure 15. Depending on the design, it is possible to recover from reprocessing the spent fuel enough fissile plutonium for the reactor's own needs, with some left over for future breeder reactors or for use in conventional reactors (see Figure 14).

Fast neutron reactors have a high thermal efficiency due to their high-temperature operation. They also have a high power density and are normally cooled by liquid metal, such as sodium, lead, or lead-bismuth, with high conductivity and boiling point and no moderating effect. They operate at around 500-550°C at or near atmospheric pressure. Although in many ways liquid metal coolant is difficult to handle chemically, in some respects it is more benign overall than very high pressure water, which requires robust engineering on account of the pressure. Experiments on a 19-year old UK breeder reactor before it was decommissioned in 1977 showed that the liquid sodium cooling system made it less sensitive to coolant failures than the more conventional very high pressure water and steam systems in light water reactors. More recent operating experience with large French and UK prototypes has confirmed this.

About 20 FBRs have already been operating, some since the 1950s.

The fast breeder reactor has the potential for utilizing virtually all of the uranium produced from mining operations. As

---

[9] Both U-238 and Pu-240 are "fertile" (materials), i.e. by capturing a neutron they become (directly or indirectly) fissile Pu-239 and Pu-241 respectively.

Table 9: Fast breeder reactors

		Output MWe gross	MW (thermal)	Full operation
USA	EBR 1	0.2		1951-63
	EBR 2	20		1963-94
	Fermi 1	66		1963-72
	SEFOR	20		1969-72
	Fast Flux Test Facility		400	1980-93
UK	Dounreay FR	15		1959-77
	Prototype FR	270		1974-94
France	Rapsodie		40	1966-82
	Phenix *	250		1973-
	Superphenix 1	1240		1985-98
Germany	KNK 2	21		1977-91
India	FBTR		40	1985-
Japan	Joyo		140	1978-
	Monju	280		1994-96[10]
Kazakhstan	BN 350*	135		1972-99
Russia	BR 5 /10		5 /10	1959-71, 1973-
	BOR 60	12		1969-
	BN 600*	600		1980-

Units in commercial operation
Source: OECD NEA 1997, Management of Separated Plutonium. (updated)

described in section 3.2, about 60 times more energy overall can be extracted from the original uranium by the fast breeder cycle than can be produced by the current light water reactors operating in "open cycle". This extremely high energy efficiency makes the breeder an attractive energy conversion system. However, high capital costs and an abundance of relatively low cost uranium means that they are generally not competitive at present.

For this reason, interest in fast reactors was very low until recently. The 1250 MWe French Superphenix FBR operated from 1985 to 1998 before being closed by political edict. The first of four 500 MWe Indian FBRs is under construction, to pave the way to greater use of thorium as a fuel, at Kalpakkam. Japan's Monju prototype commercial FBR was connected to the grid in August 1995 but was then shut down due to a major sodium leak. It is due to restart about 2007.

The Russian BN-600 fast breeder reactor has been supplying electricity to the grid since 1981 and is said to have the best operating and production record of all Russia's nuclear power units. The BN-350 FBR operated in Kazakhstan for 27 years, and about half of its output was used for water desalination. The BN-800 fast reactor being built at Beloyarsk is designed to supersede the BN-600 unit there and utilize

[10] Could restart 2008

Figure 15: Fission in conventional and fast neutron reactors

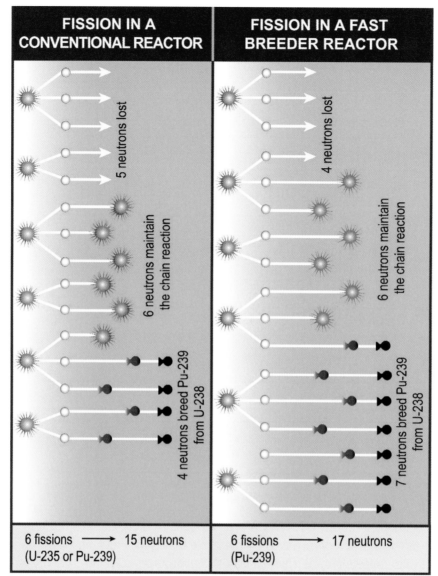

Contrast between conventional ("thermal") reactor and fast neutron reactor showing how typically more neutrons are produced in the fast reactor (17 instead of 15 from 6 fissions), thus enabling the system to breed more fissile material than is consumed if desired. In this example 4 neutrons are available for breeding Pu-239 in the conventional reactor, but 7 are available in the fast reactor. The exact numbers involved will depend on design and operation.

MOX fuel with both reactor-grade and weapons plutonium. Further BN-800 units are planned. This represents a technological advantage for Russia and has significant export or collaborative potential with Japan.

There is renewed interest in fast reactors due to their ability to fission actinides, including those which may be recovered from ordinary reactor used fuel. The fast neutron environment minimizes neutron capture reactions and maximizes fissions in actinides. This means less long-lived nuclides in high-level wastes (the fission products being preferable due to their shorter lives).

See also: relevant sections of WNA information papers on Advanced Nuclear Power Reactors, Small Nuclear Power Reactors and Fast Neutron Reactors.

4.6 VERY SMALL NUCLEAR POWER PLANTS

Two small nuclear power reactors with prospects of deployment in the next 10-15 years are noteworthy. They are being developed cooperatively. Both are fast-neutron reactors built in a factory with a sealed core and shipped to site where they would be installed below ground level. They operate at atmospheric pressure and their passive safety features and automatic load following is achieved due to the reactivity feedback – constrained coolant flow leads to higher core temperature, which slows the reaction.

The Super-Safe, Small & Simple – **4S "nuclear battery"** – system is being developed by Toshiba in Japan. It uses well-proven sodium as coolant (with electromagnetic pumps) and has passive safety features. The unit would drive a steam cycle and be capable of three decades of continuous operation without refuelling. Metallic alloy fuel is enriched to less than 20% U-235. Steady power output over the core lifetime is achieved by progressively moving upwards an annular reflector around the slender core (0.7 m diameter, 2 m high). After 14 years a neutron absorber at the centre of the core is removed and the reflector repeats its slow movement up the core for 16 more years. In the event of power loss the reflector falls to the bottom of the reactor vessel, slowing the reaction, and external air circulation gives decay heat removal.

Overnight plant cost for a 10 MWe version is projected at US$ 2500/kW and power cost at 5-7 cents/kWh, which is very competitive with diesel fuel in many locations. The design has gained considerable support in Alaska, and toward the end of 2004 the town of Galena granted initial approval for Toshiba to build a 4S reactor in that remote location. A pre-application licensing review is being sought with a view to a demonstration unit operating by 2012. Its design is sufficiently similar to an earlier liquid metal-cooled inherently-safe reactor design, which went part-way through the US approval process for it to have good prospects of licensing.

The Secure Transportable Autonomous Reactor – **STAR** – is a lead-cooled modular reactor with passive safety features. Its primary coolant circulates by natural convection. A smaller variant is the Small Sealed Transportable Autonomous Reactor – **SSTAR**, being developed by Argonne in the USA. It has lead or Pb-Bi cooling, runs at 566°C and has an integral steam generator inside the sealed unit. After a 20-year life without refuelling, the whole reactor unit is then returned for recycling the fuel. The core is 1 m in diameter and 0.8 m high. The main development is now focused on a 20 MWe version and a demonstration unit is envisaged about 2015.

4.7 THORIUM CYCLE

Near-breeder or thorium cycle reactors are similar to fast breeders in that a fertile material around the core – naturally-occurring thorium (Th-232) – will absorb slow neutrons to become (indirectly) fissile uranium-233. This will produce a chain reaction yielding heat while surplus neutrons convert more thorium to U-233.

The technology is considered by some to be attractive because plutonium (and other long-lived transuranic elements) production is avoided, abundant thorium is used as a fuel, and the efficiency of fuel use approaches that of the fast breeder reactor. However, the amount of fissile uranium produced is not quite enough to sustain the reaction, hence the term "near-breeder" is generally used, and some input of fissile U-235 or Pu-239 will always be required. Also, it is necessary to reprocess the used fuel to recover the U-233 and recycle it, but such reprocessing of thorium fuel has not yet been done on any scale.

Though a focus of interest for 30 years, only India is actively developing the concept via a three-stage programme which will culminate in the Advanced Heavy Water Reactors briefly described previously (section 4.3).

THE "BACK END" OF THE NUCLEAR FUEL CYCLE

5.1 NUCLEAR "WASTES"

Despite its demonstrable safety record over half a century, one of the most controversial aspects of the nuclear fuel cycle today is the question of management and disposal of radioactive wastes.

The most difficult of these are the high-level wastes, and there are two alternative strategies for managing them:

- Reprocessing used fuel to separate them (followed by vitrification and disposal)
- Direct disposal of the used fuel containing high levels of radioactivity as waste

The principal nuclear wastes remain locked up securely in the ceramic reactor fuel or glass.

As outlined in Chapters 3 and 4, "burning" the fuel of the reactor core produces fission products, such as various isotopes of barium, strontium, caesium, iodine, krypton and xenon (Ba, Sr, Cs, I, Kr, and Xe). Many of the isotopes formed as fission products within the fuel are highly radioactive and correspondingly short-lived.

As well as these smaller atoms formed from the fissile portion of the fuel, various transuranic isotopes are formed by neutron capture. These include Pu-239, Pu-240 and Pu-241[1], as well as others arising from some of the U-238 in the reactor core by neutron capture and subsequent beta decay. All are radioactive and apart from much of the fissile plutonium, which is "burned", they remain within the used fuel when it is removed from the reactor. The transuranic isotopes and other actinides[2] form most of the long-lived portion of high-level waste.

Today there is renewed interest in reprocessing used fuel, both to recover usable uranium and plutonium and to recover the long-lived transuranics so that the remaining high-level waste is more easily disposed of due to its shorter-lived radioactivity. This interest is coupled with the prospect of much greater use of fast reactors after 2020, which have the best capacity to fission such actinides[3].

While the civil nuclear fuel cycle generates various wastes – many of them potentially hazardous – these do not become pollution, since virtually all are contained and managed. In fact, nuclear power is the only major energy-producing industry which takes full responsibility for all its wastes and fully costs this into the product. Furthermore, the expertise developed in managing civil wastes is now starting to be applied to military wastes, which pose a real environmental problem or threat in a few parts of the world.

Radioactive wastes comprise a variety of materials requiring different types of management to protect people and the environment. Management and disposal is technically straightforward.

Radioactive wastes are normally classified as low-level, intermediate-level or high-level wastes, according to the amount and types of radioactivity in them. In a country such as the UK, radioactive wastes comprise about 1% of all toxic wastes which require management and disposal.

Another factor in managing radioactive wastes is the time that they are likely to remain hazardous. This depends on the kinds of radioactive isotopes in them, and particularly the half-lives characteristic of each of those isotopes. The half-life is the time it takes for a

---

[1] It is Pu-241 which decays to give us the americium-241 used in household smoke detectors.
[2] Actinides are elements with atomic number of 89 (actinium) or above; transuranics are those above 92 (uranium).
[3] The fast neutron environment minimizes neutron capture reactions and maximizes fissions in actinides.

given radioactive isotope to lose half of its radioactivity. After four half-lives the level of radioactivity is $1/16$ th of the original and after eight half-lives, $1/256$ th.

The various radioactive isotopes arising from nuclear energy have half-lives ranging from fractions of a second to minutes, hours or days, through to billions of years. Radioactivity decreases with time as these isotopes decay into stable, non-radioactive ones. The rate of decay of an isotope is inversely proportional to its half-life; a short half-life means that it decays rapidly. Hence, for each kind of radiation, the higher the intensity of radioactivity in a given amount of material, the shorter the half-lives involved.

Three general principles are employed in the management of radioactive wastes:
- Concentrate-and-contain
- Dilute-and-disperse
- Delay-and-decay

The first two are also used in the management of non-radioactive wastes. The wastes are either concentrated and then isolated, or very small quantities are diluted to acceptable levels (often with a delay to allow decay) and then discharged to the environment. Delay-and-decay however is unique to radioactive waste management; it means that the waste is stored and its radioactivity is allowed to decrease naturally through decay of the radioisotopes in it. Then it can proceed to disposal.

In the civil nuclear fuel cycle the main focus of attention is high-level waste containing the fission products and transuranic elements formed in the reactor core.

The **high-level waste** may be used fuel itself, or the principal waste arising from reprocessing this fuel. Either way, the volume is modest – about 25-30 tonnes of used fuel or three cubic metres per year of vitrified waste for a typical large nuclear reactor (1000 MWe, light water type). This can be effectively and economically isolated. Its level of radioactivity falls rapidly (see Figure 18, though the curve shown starts at one year). For instance, a newly discharged light water reactor fuel assembly is so radioactive that it emits several hundred kilowatts of heat, but after a year this is down to 5 kW and after five years, to 1 kW. In 40 years the radioactivity in it drops to about one thousandth of the level at discharge.

If the used fuel is reprocessed, the 3.5% of it which emerges as separated high-level waste is largely liquid, containing the "ash" from burning uranium. It consists of the highly-radioactive fission products and some heavy elements with long-lived radioactivity. It generates a considerable amount of heat and requires cooling. This is dried and vitrified into borosilicate glass (similar to Pyrex) for encapsulation, interim storage, and eventual disposal deep underground. This is the policy adopted by the UK, France, Germany, Switzerland, Japan, China and India (see sections 5.2 and 5.3).

On the other hand, if used reactor fuel is not reprocessed, all the highly radioactive fission product isotopes and the much smaller quantity of long-lived actinides remain in it, and so whole fuel assemblies are treated as high-level waste. The direct disposal option is being pursued by the USA, Canada, Finland and Sweden (see section 5.4).

A number of countries have deferred choosing between reprocessing and direct disposal.

High-level wastes make up only 3% of the volume of all radioactive wastes worldwide, but they hold 95% of the total radioactivity in them.

Figure 16: What happens in a light water reactor over 3 years

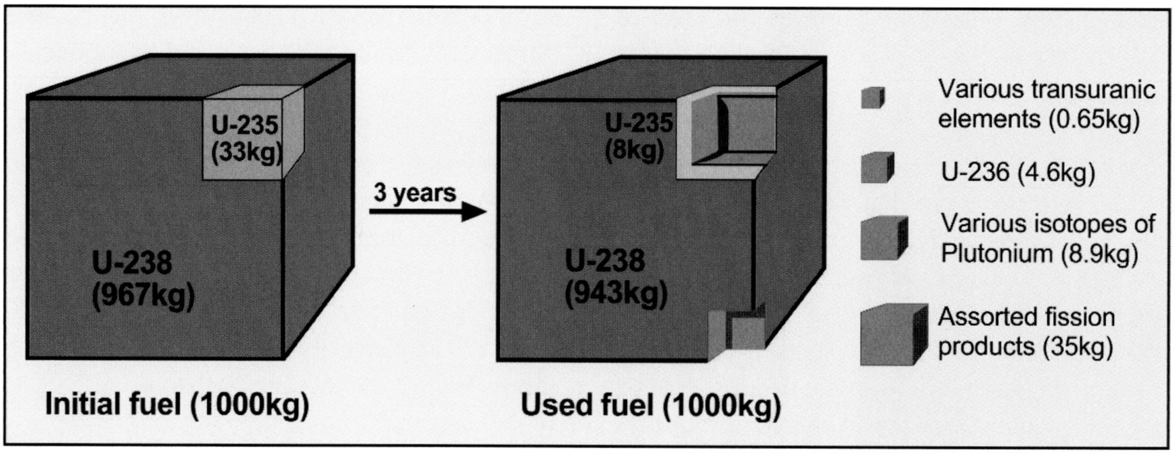

Source: Scientific American, June 1977.

In addition to the high-level wastes from nuclear power production, all use of radioactive materials in hospitals, laboratories and industry generates what are termed low-level wastes (cleaning equipment, gloves, clothing, tools, etc.), which are not dangerous to handle but must be disposed of more carefully than normal garbage. **Low-level wastes** come from hospitals, universities and industry, as well as the nuclear power industry. They may be incinerated. Ultimately they are usually buried in shallow landfill sites. Provided all highly toxic materials are first separated and included with intermediate-level wastes, this has been shown to be an effective means of waste management for such relatively innocuous materials. Many countries have final repositories in operation for low-level wastes, which have about the same level of radioactivity as a low-grade uranium orebody. They amount to over fifty times the volume of the annual arisings of high-level wastes. Worldwide they make up 90% of the volume but have only 1% of the total radioactivity of all radioactive wastes.

Some low-level liquid wastes from reprocessing plants are discharged to the sea. These include radionuclides which are distinctive, notably technetium-99 (sometimes used as a tracer in environmental studies), and this can be discerned many hundred kilometres away. However, such discharges are regulated and controlled, and the maximum radiation dose anyone receives from them is a small fraction of natural background.

Nuclear power stations and reprocessing plants release small quantities of radioactive gases (e.g. krypton-85 and xenon-133) and trace amounts of iodine-131 to the atmosphere. However, they have short half-lives, and the radioactivity in the emissions is diminished by delaying their release. Also, the first two are chemically inert. The net effect is too small to warrant consideration in any life-cycle analysis.

Intermediate-level wastes mostly come from the nuclear industry and research reactors. They are more radioactive and need to be shielded from people before treatment and disposal. They typically comprise resins, chemical sludges and reactor components, as well as contaminated materials from reactor decommissioning. Mostly these wastes are embedded in concrete for disposal. Generally short-lived waste (mainly from reactors) is buried, but long-lived waste (from reprocessing nuclear fuel) will be disposed of deep underground. Worldwide it makes up 7% of the volume of radioactive wastes and has 4% of the radioactivity.

5.2 REPROCESSING USED FUEL

The principal reason for reprocessing is to recover unused uranium and plutonium in the discharged fuel elements. A secondary reason is to reduce the volume and/or radioactivity of material to be disposed of as high-level waste.

Reprocessing avoids the waste of a valuable resource because most of the used fuel (uranium at less than 1% U-235 and a little plutonium) can be recycled as fresh fuel elements, saving some 30% of the natural uranium otherwise required. The uranium and plutonium become MOX fuel, and are a significant resource. The remaining radioactive high-level wastes are then converted into compact, stable, insoluble solids for disposal, which is easier than disposing of the more bulky used fuel assemblies. In future reprocessing is likely also to remove the long-lived transuranic elements (to be burned in a reactor), leaving only shorter-lived fission products as the waste and further simplifying disposal.

A 1000 MWe light water reactor produces about 27 tonnes of used fuel per year. So far, more than 90,000 tonnes of used fuel from commercial power reactors has been reprocessed, and annual capacity is now some 5000 tonnes per year.

Used fuel assemblies removed from a reactor are very radioactive and produce heat. They are therefore put into large tanks or "ponds" of water for cooling, while the three metres of water over them shields the radiation. Here they remain, either at the reactor site or at the reprocessing plant, for a number of years as the level of radioactivity decreases considerably. For most types of fuel, reprocessing occurs about five years after reactor discharge.

Used fuel may be transported after initial cooling, using special shielded casks which hold only a few (e.g. six) tonnes of used fuel but weigh about 100 tonnes (see box *Transporting Radioactive Materials* below). Transport of used fuel and other high-level waste is tightly regulated.

Reprocessing of used oxide fuel involves dissolving the fuel elements in nitric acid. Chemical separation of uranium and plutonium is then undertaken. The Pu and U can be returned to the input side of the fuel cycle – the plutonium straight to fuel fabrication and the uranium to the conversion plant prior to re-enrichment, though in fact most is put into long-term storage. Figure 13 shows reprocessing and MOX fuel fabrication on opposite sides of the diagram, but for recycled fuel they may be on a single site. The remaining liquid after Pu and U are removed is high-level waste, containing about 3.5% of the used fuel. It is highly radioactive and continues to generate a lot of heat.

Table 10: World commercial reprocessing capacity

Light water reactor fuel:	France, La Hague	1700
	UK, Sellafield (THORP)	900
	Russia, Ozersk (Mayak)	400
	Japan	40
	Total approx	**3040** tonnes per year
Other nuclear fuels:	UK, Sellafield	1500
	India	275
	Total approx	**1750** tonnes per year

Sources: OECD/NEA Nuclear Energy Data 2005, Nuclear Eng. International Handbook 2005.

Future development of reprocessing is likely to separate uranium for eventual recycle, and plutonium together with minor actinides (transuranic elements) for immediate recycle, leaving only fission products in the waste stream. A further development would be then to separate some, especially longer-lived, fission products for transmutation.[4]

A great deal of reprocessing has been going on since the 1940s, mainly for military purposes, to recover plutonium (from low burnup fuel) for weapons. In the UK, metal fuel elements from the first generation gas-cooled commercial reactors have been reprocessed at Sellafield for about 40 years. The 1500 t/yr plant has been successfully developed to keep abreast of evolving safety, hygiene and other regulatory standards. From 1969 to 1973 oxide fuels were also reprocessed, using part of the plant modified for the purpose. A new 1200 t/yr thermal oxide reprocessing plant (THORP) was commissioned in 1994 but has never run at full capacity.

In the USA, three plants for the reprocessing of civilian oxide fuels have been built, but for technical, economic and political reasons all were shut down or aborted. In all the USA has over 250 plant-years of reprocessing operational experience, the vast majority being at government-operated defence plants since the 1940s.

In France one 400 t/yr reprocessing plant operated for metal fuels from early gas-cooled reactors at Marcoule. At La Hague, reprocessing of oxide fuels has been carried out since 1976, and two 800 t/yr plants are now operating. India has a 100 t/yr oxide fuel plant operating at Tarapur with others at Kalpakkam and Trombay, and Japan is commissioning a major plant at Rokkasho, having had most of its used fuel reprocessed in Europe meanwhile. It

has had a small (100 t/yr) plant operating. Russia has a 400 t/yr oxide fuel reprocessing plant at Ozersk and is planning to upgrade this.

See also: WNA information paper Processing of Nuclear Wastes, which includes new reprocessing technologies.

After reprocessing, the recovered uranium requires re-enrichment, so it goes first to a conversion plant. This is complicated by the presence of impurities and two new isotopes in particular, U-232 and U-236, which are formed by neutron capture in the reactor. Both decay much more rapidly than U-235 and U-238, and one of the daughter products of U-232 emits very strong gamma radiation, which means that shielding is necessary in the enrichment plant. U-236 is a neutron absorber, which impedes the chain reaction and means that a higher level of U-235 enrichment is required in the product to compensate. Being lighter, both isotopes tend to concentrate in the enriched (rather than depleted) output, so reprocessed uranium which is re-enriched for fuel must be segregated from enriched fresh uranium.

Separated plutonium is recycled via a dedicated MOX fuel fabrication plant, which will often be integrated with the reprocessing plant which separated it. In France the reprocessing output is coordinated with MOX plant input, to avoid building up stocks of plutonium. (If plutonium is stored for some years the level of americium-241 – the isotope used in household smoke detectors – will accumulate and make it difficult to handle through a MOX plant due to the elevated levels of gamma radioactivity.)

See also: WNA information paper Mixed Oxide Fuel.

---

[4] Notably iodine (I-129), technetium (Tc-99), caesium (Cs-135) and strontium (Sr-90).

Table 11: World mixed oxide fuel fabrication capacities (tonnes per year)

	2005	2010
Belgium	35	0
France	145	195
Japan	10	140
Russia	-	?
UK	40	40
Total for LWR	230	375

Source: OECD/NEA 2005 Nuclear Energy Data, Nuclear Eng. International handbook 2005.

5.3 HIGH-LEVEL WASTES FROM REPROCESSING

Despite the small quantities involved (see section 5.1), high-level wastes from reprocessing used reactor fuel require great care in handling, storage and disposal because they comprise fission products and transuranic elements which emit alpha, beta and gamma radiation at high levels, as well as a lot of heat. The heat arises mainly from the fission products, which mostly have the shorter half-lives. These are the materials popularly thought of as "nuclear wastes". In future – some years off – these wastes may comprise mainly fission products, which means that in a hundred-year perspective they are no different from today's separated wastes; but taking a thousand-year perspective, they are much less radioactive.

It is worth noting that wastes from weapons programmes will continue to overshadow civil nuclear wastes in countries like the USA and Russia for many decades, no matter how rapidly commercial nuclear power expands. The legacy of these wastes, dating from the 1940s, in polluted land, leaking storage tanks and the prospect of very high clean-up costs remains with those countries which produced them.

The liquid wastes generated in reprocessing plants are stored temporarily in cooled multiple-walled stainless steel tanks surrounded by reinforced concrete. These need to be changed into compact, chemically inert solids before considering the question of permanent disposal. The main method of solidifying liquid wastes is vitrification. The Australian Synroc (synthetic rock) is a more sophisticated way to immobilize such waste, but this has not yet been commercially developed for civil wastes.

Commercial vitrification plants are based on calcination of the wastes (evaporation to a dry powder), followed by incorporation in borosilicate glass. The molten glass is mixed with the dry wastes and poured into large stainless steel canisters, each holding 400 kg. A lid is then welded on. A year's waste from a 1000 MWe reactor is contained in 5 tonnes of such glass, or about twelve canisters, each 1.3 metres high and 0.4 metres diameter. In the UK these are stored vertically in silos, ten deep.

The 90,000 tonnes of used fuel so far reprocessed worldwide will have been reduced to about 6000 cubic metres of vitrified high-level wastes – a 20 m x 20 m pile, 16 m high in visual terms (see also Figure 17).

Processes such as these have been developed and tested in pilot plants since the 1960s. In the UK at Harwell several tonnes of high-level wastes from reprocessed fuel were vitrified by 1966, but research was then set aside until there were enough such wastes to give the issue a higher priority. High-temperature

Borosilicate glass from the first waste vitrification plant in UK in the 1960s. This block contains material chemically identical to high-level waste from reprocessing used fuel. A piece this size from modern vitrification plants would contain the total high-level waste arising from nuclear electricity generation for one person throughout a lifetime.

leaching tests on this glass showed that it has remained insoluble even where some physical breakdown of the glass had occurred. Similar results have been obtained on French wastes vitrified between 1969 and 1972.

Vitrification of civil high-level radioactive wastes first took place on an industrial scale in France in 1978. It is now carried out commercially at five facilities in Belgium, France and the UK with capacity of 2500 canisters (1000 tonnes) per year.

In 1996 two vitrification plants were opened in the USA. One, at West Valley, NY, was to treat 2.2 million litres of high-level waste from civil nuclear fuel reprocessed there 25 years earlier, and the other was at Savannah River, SC, to vitrify a larger quantity of military waste.

Vitrified wastes will be stored for some time before final disposal, to allow heat and radioactivity to diminish. In general, the longer the material can be left before disposal the

easier it is to handle and the less space is required in a repository. Depending on the actual disposal methods adopted, there will be some 50 years between reactor and disposal.

All handling of such materials involves the use of protective shielding and procedures to ensure the safety of people involved. As in all situations where gamma radiation is involved, the simplest and cheapest protection is distance – ten times the distance reduces exposure to one percent – or mass.

When separated high-level wastes (or used fuel assemblies) are moved from one place to another, robust shipping containers are used. These are designed to withstand all credible accident conditions without leakage or reduction in their radiation shielding effectiveness. Where such containers have been involved in serious accidents over the years, they have created no radioactivity hazard at all. The high standards of integrity designed into these containers also make them difficult to breach with explosives and therefore unattractive as an object for sabotage attempts.

Transporting Radioactive Materials

About 20 million shipments of radioactive material (which may be either a single package or a number of packages sent from one location to another at the same time) take place around the world each year. Radioactive material is not unique to the nuclear fuel cycle and most shipments of such material are not fuel cycle-related. Radioactive materials are used extensively in medicine, agriculture, research, manufacturing, non-destructive testing and minerals' exploration. Nuclear materials have been transported since before the advent of nuclear power over 40 years ago. The procedures employed are designed to ensure the protection of the public and the environment.

Nuclear fuel cycle facilities are located in various parts of the world, and materials of many kinds need to be transported between them. Many of these are similar to materials used in other industrial activities. However, the nuclear industry's fuel and waste materials are radioactive, and it is these "nuclear materials" about which there is most public concern.

Transport is an integral part of the nuclear fuel cycle. There are some 440 nuclear power reactors in operation in 30 countries, but uranium mining is viable in only a few areas. Furthermore, in the course of over 40 years of operation by the nuclear industry, a number of specialized facilities have been developed in various locations around the world to provide fuel cycle services. It is clear that there is a need to transport nuclear fuel cycle materials to and from these facilities. Indeed, most of the material used in nuclear fuel is transported several times during its "life". Transport operations are frequently international, and are often over large distances. Nuclear materials are generally transported by specialized transport companies.

Since 1971 there have been some 7000 shipments of used fuel (over 35,000 tonnes) over more than 30 million kilometres with no property damage or personal injury, no breach of containment, and very low dose rate to the personnel involved (e.g. 0.33 mSv/yr per operator at La Hague). Some 300 sea voyages have been made carrying used nuclear fuel or separated high-level waste over a distance of more than 8 million kilometres. The major company involved has transported over 4000 casks, each of about 100 tonnes, carrying 8000 tonnes of used fuel or waste. A quarter of these have been through the Panama Canal.

In Sweden, more than 80 large transport casks are shipped annually from nuclear power stations (all on the coast) to a central interim waste storage facility called CLAB. A purpose-built 2000 tonne ship is used for moving the used fuel. In the USA more than 3000 shipments of used nuclear fuel have been made over 2.7 million kilometres with no harmful release of radiation.

Packaging

The principal assurance of safety in the transport of nuclear materials is the design of the packaging, which must allow for foreseeable accidents. The consignor bears primary responsibility for this.

"Type A" packages are designed to withstand minor accidents and are used for medium-activity materials, such as medical or industrial radioisotopes. Ordinary industrial containers are used for low-activity material, such as U_3O_8.

Spent fuel cask being loaded from train on to ship

Packages for high-level waste (HLW) and used fuel are robust and very secure containers known as "Type B" packages. They also maintain shielding from gamma and neutron radiation, even under extreme conditions. There are over 150 kinds of Type B packages, and the larger ones weigh up to 110 tonnes each when empty and hold up to 6 tonnes of used fuel.

In France alone there are some 750 shipments each year of Type B packages, among 15 million shipments of goods classified as "dangerous materials", 300,000 of these being radioactive materials of some kind.

Smaller amounts of high-activity materials (including plutonium) transported by aircraft will be in "Type C" packages, which give greater protection in all respects than Type B packages in accident scenarios.

To limit the risk in handling of highly radioactive materials, dual-purpose containers (casks), which are appropriate for both storage and transport of used nuclear fuel, are often used.

Regulation

Since 1961 the International Atomic Energy Agency (IAEA) has published advisory regulations for the safe transport of radioactive material. These regulations have come to be recognized throughout the world as the uniform basis for both national and international transport safety requirements in this area. Requirements based on the IAEA regulations have been adopted in about 60 countries, as well as by the International Civil Aviation Organization (ICAO), the International Maritime Organization (IMO), and regional transport organizations.

The fundamental principle applied to the transport of radioactive material is that the protection comes from the design of the package, regardless of how the material is transported.

See also: WNA information paper Transporting Nuclear Material.

5.4 STORAGE AND DISPOSAL OF USED FUEL AS "WASTE"

The direct disposal option for used fuel is the policy of many countries, though usually it will be recoverable. While separated high-level wastes are vitrified to make them insoluble and physically stable, used fuel destined for direct disposal is already in a very stable ceramic form as UO_2.

There is about 270,000 tonnes of used fuel in storage, much of it at reactor sites. Annual arisings of used fuel are about 12,000 tonnes, and 3000 tonnes of this goes for reprocessing. Final disposal is therefore not urgent in any logistics sense.

Sweden has had since 1988 a fully operational central long-term spent fuel storage facility (CLAB) with capacity for all the country's used fuel, which is sent to it after storage at the reactor site for only a year or so. At CLAB the spent fuel is handled under water, for cooling and radiological shielding, and stored for some 40 years. By 2020 this storage will be full and a final repository should be ready.

In considering the used fuel itself or the waste extracted from it, an important feature is the rate at which it cools and radioactivity decays. Forty years after removal from the reactor less than one thousandth of its initial radioactivity remains, and it is much easier to deal with (see Figure 18). This feature sets nuclear waste apart from chemical wastes, which remain hazardous unless they are destroyed. The longer nuclear wastes are stored, the less hazardous they are and the more readily they can be handled.

In the USA and several other countries all used fuel remains stored at reactor sites by the utilities, and at present this is as far as the fuel cycle goes. In the USA it is intended that used fuel should be transferred from the reactor site storage ponds or dry cask storage to a federal repository at Yucca Mountain in Nevada. Utility customers pay a fee of 0.1 cent per kilowatt-hour for management and eventual disposal of their spent fuel. By the end of 2005 this amounted to over US$ 25 billion.

Figure 18: Decay in radioactivity of fission products in one tonne of spent PWR fuel

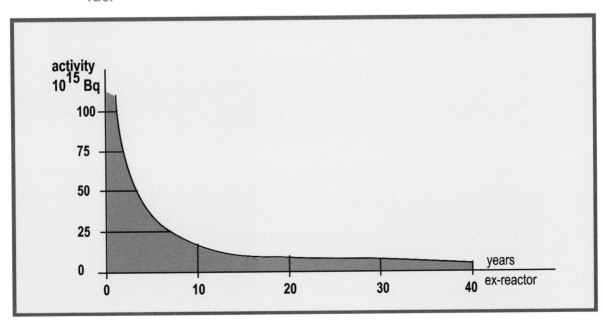

5.5 DISPOSAL OF SOLIDIFIED WASTES

Whether the final high-level waste is vitrified material from reprocessing or entire used fuel assemblies, it needs eventually to be disposed of safely. In addition to concepts of safety applied elsewhere in the nuclear fuel cycle, this means that it should not require any ongoing management after disposal, or after closure of the repository. While final disposal of high-level wastes will not take place for some years yet, preparations are being made at a rate appropriate to the nature and quantities of the wastes involved.

As part of an ongoing review of waste management strategies, the Radioactive Waste Management Committee of the OECD Nuclear Energy Agency[5] reassessed the basis for the geological disposal of radioactive waste from an environmental and ethical perspective. Considerations of intergenerational equity were emphasized. In 1995 the Committee confirmed "that the geological disposal strategy can be designed and implemented in a manner that is sensitive and responsive to fundamental ethical and environmental considerations", and concluded that:

- "it is justified, both environmentally and ethically, to continue development of geological repositories for those long-lived radioactive wastes which should be isolated from the biosphere for more than a few hundred years", and that

- step-by-step "implementation of plans for geological disposal leaves open the possibility of adaptation, in the light of scientific progress and [developing] social acceptability over several decades, and does not exclude the possibility that other options could be developed at a later stage".

The final disposal of high-level waste must be done with a very high degree of assurance. The question is how to be confident of this before it has been undertaken on a large scale over many

Figure 19: Activity of high-level waste from one tonne of spent fuel

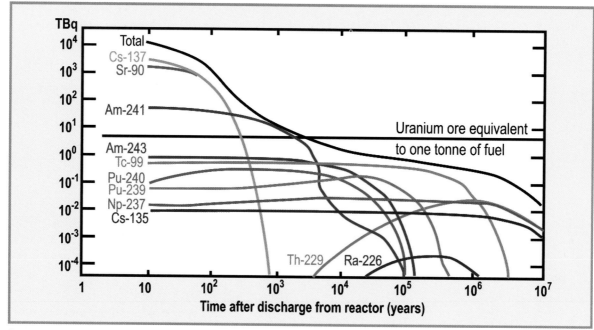

Source: IAEA, 1992 - Radioactive Waste Management

[5] The Environmental and Ethical Basis of Geological Disposal of Long-Lived Radioactive Wastes, OECD/NEA, 1995 (see Appendix 3).

years. It is apparent that a high level of confidence can in fact be achieved by continuing the careful research and design which has been now going on for some time. The problems involved are neither very large nor complicated. In most respects they are not even novel.

First, the separated waste or used fuel is in a stable and insoluble form. Second, it is (or will be) encapsulated in heavy stainless steel casks or in canisters which are corrosion-resistant (e.g. stainless steel and copper). Third, it will be geologically isolated.

The degree of hazard involved is indicated by Figure 19, and the picture would be similar for used fuel except that the plutonium lines would be higher up (around 10 Tbq). It can readily be seen that if the minor actinides americium (Am-241, Am-243), neptunium-237 and curium (alpha-decaying to plutonium) were removed in a further reprocessing stage, the radioactivity after 100 years would be much reduced.

Two important conclusions can be drawn from the changes shown in Figure 19. The first is that the radiological hazard falls by a factor of nearly a thousand between 10 and 1000 years, with relatively little change subsequently. This is because nearly all of the short half-life fission products from the chain reaction will have decayed to negligible levels, leaving behind small quantities of transuranic elements such as americium and neptunium, which generally have much longer half-lives. A thousand years is still a long time in human terms, but the object is to put it into stable geological formations where geological time becomes a more meaningful reference. Even the time needed for plutonium to decay to a low level is brief geologically.

The second important point from Figure 19 is that the relative radioactivity of the waste after 1000 years is much the same as that of the corresponding amount of uranium ore. Of course, toxic components of a uranium orebody, which outcrops at the surface of the earth, actually do find their way into the human food chain. Waste material in ceramic form buried some 500 metres below the surface in a dry, stable geological structure will have virtually no conceivable chance of doing so. (However, this is not to say that surface uranium deposits are dangerous, as the amounts which reach anybody are very small.)

Most countries with nuclear facilities planned or operational have active programmes aimed at defining and testing suitable deep geological disposal sites. The aim of this work is to locate areas where multiple barriers can be established between the wastes and the human environment. Some of the barriers, both natural and artificial, being sought are:

- Converting the waste to an insoluble form (glass, Synroc or UO_2) – see sections 5.3 and 5.4
- Sealing the waste in corrosion-resistant containers
- In wet rock, packing with bentonite clay to inhibit groundwater movement and to insulate from minor earth movement
- Locating the waste deep underground (e.g. 500 m deep) in a stable rock structure

Three types of geological structures are being widely studied for this purpose – hard crystalline rocks, clays and rock salt beds. Suitable locations have been identified in several countries and sites are now undergoing detailed evaluation. Most approaches plan to utilize conventional mining techniques involving shaft-sinking and developing extensive drives and rooms. These will provide sufficient area for the canisters to be placed on each level. One purpose-built deep geological repository is now operating in New Mexico, USA, but this is only for long-lived military wastes.

The problems involved in carrying out this work are essentially technical. Conventional mining and engineering design techniques together with monitoring of rock temperatures and stresses will enable disposal operations to be carried out to a very high order of safety. The engineering and organizational tasks of maintaining effective isolation of hazardous materials are not new, nor are they peculiar to nuclear wastes.

The question of geological stability of the rock structure is very important for the long-term integrity of the waste depository. There are a number of rock structures which have been stable for more than half of the Earth's 4500 million years, suggesting little likelihood of significant movement for isolation periods of 1000 years or more.

While deep geological disposal of nuclear wastes is potentially permanent, it is possible to leave open the option of making the material retrievable by future generations. Used fuel will always remain a resource, and it is possible that decades hence, it will have a value which makes it worth recovering for recycling.

The Japanese Cavern Retrievable (CARE) concept involves two distinct stages: ventilated underground caverns with the wastes in overpacks (hence shielded) and fully accessible, followed by backfilling and sealing the caverns after around 300 years. The initial institutional control period – during which most of the radiological decay of the wastes occurs – ensures that thermal load is much reduced by stage 2 allowing a much higher density of wastes than other disposal concepts. This is readily adaptable for used fuel, with the overpacks then being shipping casks.

Since in parts of the world, such as Europe, nuclear wastes comprise only about 1% of all toxic industrial wastes, it is relevant to compare their toxicity with that of common industrial

poisons used every day by industry and others of those wastes. Arsenic, of course, is routinely distributed to the environment as a herbicide and in treated timbers. Barium is not uncommon, and chlorine is in widespread domestic and industrial use. Then there are waste mercury compounds, PCBs, organochlorines and hexachlorobenzene which are extremely hazardous – many are also liquids. Considering the quantities available, these are arguably far more hazardous than civil nuclear wastes, which have given rise to no problems or hazards in fifty years. Radioactive wastes are treated much more conservatively than other toxic wastes in relation to risks to people and the environment. Most others do not break down naturally in a way which corresponds to the progressive decay of radioactivity, and thus they mostly have an infinite life.

There is now little question that disposal of high-level waste, when it comes of age, will be safe. The wastes, though very toxic when first produced, are small in quantity and no more hazardous in total than other more familiar materials. Nevertheless, they have come to epitomize the "not in my backyard" syndrome of modern society, where we find it easier to accept the benefits of technology and economic development while hoping someone else will grapple with any dirty, unpleasant or fearful aspects, however safe they may actually be.

While each country is responsible for disposing of its own wastes of all kinds, the possibility of international nuclear waste repositories is now being very actively considered, and Russia has enacted legislation to enable it there.

A Natural Analogue: Oklo

Although highly active wastes from modern nuclear power have not yet been buried for long enough to observe the results, this process has in fact occurred naturally in at least one location. In what is now West Africa, about 2 billion years ago, at least 17 natural nuclear reactors commenced operation in a rich deposit of uranium ore at Oklo in Gabon. Each operated at less than 100 kW thermal. At that time the concentration of U-235 in all natural uranium was some 3.7% instead of 0.7% as at present[6].

These natural chain reactions, started spontaneously and with the presence of water acting as a moderator, continued for about 2 million years before finally dying away. During this long reaction period about 5.4 tonnes of fission products as well as some 2 tonnes of plutonium together with other transuranic elements were generated at the reactor locations in the orebody. It appears that each reactor operated in pulses of about 30 minutes – interrupted when the water turned to steam thereby switching it off for a couple of hours until it cooled.

The initial radioactive products have long since decayed into stable elements but close study of the amount and location of these has shown that there was little movement of radioactive wastes during and after the nuclear reactions. Plutonium and the other transuranics remained immobile. This is remarkable in view of the fact that groundwater had ready access to the wastes and they were not in a chemically inert form (such as glass). However, waste materials do not necessarily move freely through the ground even in the presence of water because of their being adsorbed on to clays[7].

Thus the only known "test" of underground nuclear waste disposal, at Oklo, was successful over a long period in spite of the characteristics of the site. Such a water-logged, sandstone/shale structure would not be considered for disposal of modern toxic wastes, nuclear or otherwise, although the clays and bitumen present played an important part in containing the wastes.

However, the Oklo example has prompted researchers to study the mobilization of uranium dioxide in groundwaters associated with other orebodies (which have not undergone fission). This will assist in assessing the long-term safety of repositories for high-level wastes. One such international analogue study took place around the Koongarra deposit in Australia's Northern Territory.

Cost

The cost of dealing properly with wastes is important. In the USA a 0.1 cent/kWh levy to finance the disposal of used fuel had accumulated some US$ 25 billion by the end of 2005. Canadian utilities collect a fee of about 0.1 cent/kWh to finance future disposal of used fuel. In Sweden a levy of some 0.3 cents/kWh finances the country's smoothly functioning waste repository for low- and intermediate-level wastes and research on disposal of used fuel. Similar arrangements are in place in other countries, with the expectation that final disposal of all nuclear wastes will be fully funded in advance.

In summary, it is apparent that safe waste management is the norm, that disposal technology exists and that full-scale demonstration at acceptable cost will be possible in several countries by 2020.

---

[6] See also Appendix 2. U-235 decays about six times faster than U-238, whose half-life is about the same as the age of the Earth.
[7] Leaks from the military waste tanks in the USA also demonstrated the ability of clay soils to retain fission products and transuranics.

5.6 DECOMMISSIONING REACTORS

So far over 90 commercial nuclear power reactors have been decommissioned, along with over 250 research reactors and a number of fuel cycle facilities.

At the end of 2005, the IAEA reported that eight power plants had been completely decommissioned and dismantled, with the sites released for unconditional use. A further 17 had been partly dismantled and safely enclosed, 31 were being dismantled prior to eventual site release and 30 were undergoing minimum dismantling prior to long-term enclosure. The broken-up pieces from dismantling are buried along with other intermediate-level wastes[8].

The IAEA has defined three options for decommissioning, after removal of the fuel. These definitions have been internationally adopted:

- Immediate Dismantling (or Early Site Release/Decon in the USA): This option allows for the facility to be removed from regulatory control relatively soon after shutdown or termination of regulated activities. Usually, the final dismantling or decontamination activities begin within a few years, depending on the facility. Following removal from regulatory control, the site is then available for re-use.

- Safe Enclosure (or Safestor): This option postpones the final removal of controls for a longer period, usually in the order of 40 to 60 years. The facility is placed into a safe storage configuration until the eventual dismantling and decontamination activities occur.

- Entombment: This option entails placing the facility into a condition that will allow the remaining radioactive material to remain on-site without the requirement of ever removing it totally. This option usually involves reducing the size of the area where the radioactive material is located and then encasing it all in a long-lived structure such as concrete, to ensure the remaining radioactivity is finally of no concern.

Each option has its benefits and disadvantages, and national policy is likely to determine which approach is adopted. In the case of immediate dismantling with early site release, responsibility for the decommissioning is not transferred to future generations. The experience and skills of operating staff can also be utilized during the decommissioning programme. Alternatively, Safe Enclosure or Safestor allows significant reduction in residual radioactivity, thus reducing radiation hazard during the eventual dismantling. The expected improvements in mechanical techniques should also lead to a reduction in both hazards and costs.

In the case of nuclear reactors, about 99% of the radioactivity is associated with the fuel which is removed before moving to any of the three options. Apart from any surface contamination of plant, the remaining radioactivity comes from "activation products", such as steel components, which have long been exposed to neutron irradiation. Their atoms are changed into different isotopes, such as iron-55, cobalt-60, nickel-63 and carbon-14.

Decommissioning of the Windscale Advanced Gas-Cooled Reactor (WAGR), UK – removal of heat exchangers

Photograph supplied by UKAEA

---

[8] Many nuclear submarines have also been decommissioned over the last decade. In the USA, after defuelling, the reactor compartments are cut out of the vessels and are transported inland to Hanford, in the state of Washington, to be buried as low-level waste.

The first two are highly radioactive, emitting gamma rays. However, their half-life is such that after 50 years from shutdown their radioactivity is much diminished and the risk to workers largely gone. Overall, in 100 years after shutdown, the level of radioactivity falls by a factor of 100,000.

Examples

To decommission its retired gas-cooled reactors at the Chinon, Bugey and St Laurent nuclear power stations, Electricité de France chose partial dismantling and Safestor, postponing final dismantling and demolition for 50 years. As other reactors will continue to operate at those sites, monitoring and surveillance do not add to the cost.

Germany chose more rapid direct dismantling over Safe Enclosure for the closed Greifswald nuclear power station in former East Germany, where five reactors had been operating. Similarly, the site of the 100 MWe Niederaichbach nuclear power plant in Bavaria was declared fit for unrestricted agricultural use in mid 1995. Following removal of all nuclear systems, the radiation shield and some activated materials, the remainder of the plant was below accepted limits for radioactivity and the state government approved final demolition and clearance of the site.

Experience in the USA has varied, but 14 power reactors are using the Safestor approach, while 10 are using, or have undertaken, immediate dismantling. Procedures are set by the Nuclear Regulatory Commission (NRC).

For Trojan nuclear power plant (1180 MWe, PWR) in Oregon, the dismantling was undertaken by the utility itself. The plant closed in 1993, steam generators were removed, transported and disposed of at Hanford in 1995, and the reactor vessel was removed and transported to Hanford in 1999. Except for the used fuel storage, the site was

Photograph supplied by UKAEA

Dismantling of WAGR using remote plasma arc cutting

released for unrestricted use in 2005.

At multi-unit nuclear power stations, the choice has been to place the first closed unit into Safestor until the others end their operating lives, so that all can be decommissioned in sequence. This will optimize the use of staff and the specialized equipment required for cutting and remote operations, and achieve cost benefits.

Thus, after 14 years of comprehensive clean-up activities, including the removal of fuel, debris and water from the 1979 accident, Three Mile Island 2 was placed in Post-Defuelling Monitored Storage (Safestor) until the operating licence of Unit 1 expires in 2014, so that both units are dismantled together. Safestor was also being used for San Onofre 1, which closed in 1992, until licences for Units 2 and 3 expired in 2013; however, after NRC changes, dismantling was brought forward to 1999, so it became a Decon project.

A US immediate dismantling project was the 60 MWe Shippingport reactor, which operated commercially from 1957 to 1982. It was used to demonstrate the safe and cost-effective dismantling of a commercial-scale nuclear power plant and the early release of the site. Defuelling was completed in two years, and five years later in 1989 the site was released for use without any restrictions. Because of its size, the pressure vessel could be removed and disposed of intact. For larger units, such components will have to be cut up.

Immediate dismantling was also the option chosen for Fort St Vrain, a 330 MWe high temperature gas-cooled reactor, which was also closed in 1989. This took place on a fixed-price contract for US$ 195 million (hence costing less than 1 cent/kWh despite a short operating life), and the project proceeded on schedule, allowing the site to be cleared and the licence to be relinquished early in 1997 – the first large US power reactor to achieve this.

Another such US Decon project was Maine Yankee, a 860 MWe plant which closed down in 1996 after 24 years' operation. The containment structure was finally demolished in 2004, and except for 5 hectares occupied by the dry store for used fuel, the site was released for unrestricted public use in 2005, on budget and on schedule.

Costs

The total cost of decommissioning is dependent on the sequence and timing of the various stages of the programme. Deferment of a stage tends to reduce its cost due to decreasing radioactivity, but this is offset by increased storage and surveillance costs.

Even allowing for uncertainties in cost estimates and applicable discount rates, decommissioning contributes less than 5% to total electricity generation costs. In the USA many utilities have revised their cost projections downwards in the light of experience, and estimates now average $325 million per reactor all-up (1998 $). Financing methods vary from country to country. Among the most common are:

- External sinking fund (Nuclear Power Levy): This is built up over the years from a percentage of the electricity rates charged to consumers. Proceeds are placed in a trust fund outside the utility's control. This is the main US system and variants are widely used. It means that sufficient funds are set aside during the reactor's operating

lifetime to cover the cost of decommissioning.

- Prepayment: Money is deposited in a separate account to cover decommissioning costs even before the plant begins operation. This may be done in a number of ways but the funds cannot be withdrawn other than for decommissioning purposes.

- Surety fund, letter of credit, or insurance: These are purchased by the utility to guarantee that decommissioning costs will be covered even if the utility defaults.

In the USA, utilities are collecting 0.1 to 0.2 cents/kWh to fund decommissioning. They must then report regularly to the NRC on the status of their decommissioning funds. As of 2001, $23.7 billion of the total estimated cost of decommissioning all US nuclear power plants had been collected, leaving a liability of about $11.6 billion to be covered over the operating lives of 104 reactors.

An OECD survey published in 2003 reported US dollar (2001) costs of decommissioning by reactor type. For western PWRs, most were $200-500/kWe; for Russian light water pressurized reactors (VVERs), costs were around $330/kWe; for BWRs, $300-550/kWe; for CANDU, $270-430/kWe. For gas-cooled reactors the costs were much higher due to the greater amount of radioactive materials involved, reaching $2600/kWe for some UK Magnox reactors.

Preparations to lift the reactor top biological shield during decommissioning at WAGR

Photograph supplied by UKAEA

6 OTHER NUCLEAR ENERGY APPLICATIONS:

HYDROGEN FOR TRANSPORT
DESALINATION
SHIPS
SPACE
RESEARCH REACTORS FOR RADIOISOTOPES

6.1 TRANSPORT AND THE HYDROGEN ECONOMY

Nuclear power is relevant to road transport and motor vehicles in two respects:

- Hybrid vehicles potentially use power from the grid for recharging.
- Hydrogen for oil refining and for fuel cell vehicles may be made electrolytically, and in the future, thermochemically using high-temperature reactors.

Hydrogen is already a significant chemical product, chiefly used in making nitrogen fertilizers and, increasingly, to convert low-grade crude oils into transport fuels[1]. Some is used for other chemical processes. World consumption is 50 million tonnes per year, growing at about 10% per annum. There is a lot of experience handling it on a large scale. Virtually all hydrogen is made from natural gas, giving rise to quantities of carbon dioxide emissions – each tonne produced gives rise to 11 tonnes of CO_2.

Like electricity, hydrogen is an energy carrier (but not a primary energy source). As oil becomes more expensive, hydrogen may replace it as a transport fuel and in other applications. This development becomes more likely as fuel cells are developed, with hydrogen as the preferred fuel. If gas also becomes expensive, or constraints are put on carbon dioxide emissions, non-fossil sources of hydrogen will become necessary.

Hydrogen itself is likely to be an important future fuel.

Like electricity, hydrogen for transport use will tend to be produced near where it is to be used. This will have major geo-political implications as industrialized countries become less dependent on oil and gas from distant parts of the world.

In the short term, hydrogen can be produced economically by electrolysis of water in off-peak periods, enabling much greater utilization of nuclear plants. In future, a major possibility is direct use of heat from nuclear energy, using a chemical process enabled by high-temperature reactors.

In the USA, producing 11 Mt/yr of hydrogen with a thermal energy of 48 GWt consumes 5% of US natural gas usage. The use of hydrogen for all US transport would require some 200 Mt/yr of hydrogen[2].

All this points to the fact that while a growing hydrogen economy already exists, linked to the worldwide chemical and refining industry, a much greater one is in sight. With new uses for hydrogen as a fuel, the primary energy demand for its production may approach that for electricity production.

Nuclear power already produces electricity as a major energy carrier. It is well placed to produce hydrogen if this becomes a major energy carrier also.

Nuclear energy can be used to make hydrogen electrolytically, and in the future, high-temperature reactors are likely to be used for thermochemical production.

The evolution of nuclear energy's role in hydrogen production over perhaps three decades is seen to be:

- Electrolysis of water, using off-peak capacity
- Use of nuclear heat to assist steam reforming of natural gas, which is energy intensive and requires temperatures of up to 900°C. This well-established process however

---

[1] For example $(CH)_n$ tar sands or $(CH_{1.5})_n$ heavy crude to $(CH_2)_n$ transport fuel
[2] 89.88 g hydrogen occupies 1 m^3 at STP. 1 tonne hydrogen occupies 11,126 m^3 at STP.

has carbon dioxide as a waste product.

- High-temperature electrolysis of steam at over 800°C, using heat and electricity from nuclear reactors. In 2004 there was demonstration of this at laboratory scale in the USA.
- High-temperature thermochemical production using nuclear heat. Several direct thermochemical processes are being developed for producing hydrogen from water. For economic production (small plant, low capital), high temperatures are required to ensure rapid throughput and high conversion efficiencies.

Efficiency of the whole process (heat to hydrogen) then moves from about 25% with today's reactors driving electrolysis, to 36% with more efficient reactors doing so, to 45% for high-temperature electrolysis of steam, to about 50% or more with direct thermochemical production.

Hydrogen from Nuclear Heat

In each of the leading thermochemical processes the high-temperature (800-1000°C), low-pressure endothermic (heat absorbing) decomposition of sulphuric acid produces oxygen and sulphur dioxide:

$$H_2SO_4 \Rightarrow H_2O + SO_2 + \frac{1}{2}O_2$$

There are then several possibilities. In the iodine-sulphur (IS) process iodine combines with the SO_2 and water to produce hydrogen iodide. This is the Bunsen reaction and is exothermic, occurring at low temperature (120°C):

$$\mathbf{I_2 + SO_2 + 2H_2O \Rightarrow 2HI + H_2SO_4}$$

The HI then dissociates to hydrogen and iodine at about 350°C, endothermically:

$$2HI \Rightarrow H_2 + I_2$$

This can deliver hydrogen at high pressure.

The net reaction is thus:

$$H_2O \Rightarrow H_2 + 1/2O_2$$

All the reagents other than water are recycled – there are no effluents.

The Japan Atomic Energy Authority (JAEA) has demonstrated laboratory-scale and bench-scale hydrogen production with the IS process, at a rate of up to 30 l/hr.

The Sandia National Laboratory in the USA and the French Commissariat à l'Energie Atomique (CEA) are also developing the IS process with a view to using high-temperature reactors for it.

General Atomics' preliminary laboratory work on thermochemical production should be complete by 2006. A 10 MW pilot hydrogen plant using fossil heat would then be built, followed by nuclear thermochemical production by 2015.

The economics of hydrogen production depend on the efficiency of the method used. The IS cycle coupled to a modular high temperature reactor is expected to produce hydrogen at $1.50 to $2.00 per kg. The oxygen by-product also has value.

For thermochemical processes an overall efficiency of greater than 50% is projected. Combined cycle plants producing both H_2 and electricity may reach efficiencies of 60%.

Production Reactor Requirements

High temperature, 750°C-1000°C, is required, though at 1000°C the conversion efficiency is three times that at 750°C. The chemical plant needs to be isolated from the nearby reactor, for safety reasons, possibly using an intermediate helium or molten fluoride loop.

Three potentially suitable reactor concepts have been identified:

- (HTGR), either the pebble bed or hexagonal fuel block type, with helium coolant at high pressure. Modules of up to 285 MWe will operate at 950°C but can be hotter.

- Advanced high-temperature reactor (AHTR), a modular reactor using a coated-particle graphite-matrix fuel and with molten fluoride salt as primary coolant[3]. This is similar to the HTGR but operates at low pressure (< 1 atmosphere) and higher temperature, and gives better heat transfer. Sizes of 1000 MWe/2000 MWt are envisaged.

- Lead-cooled fast reactor, though these operate at lower temperatures than the HTGRs – the best developed is the Russian BREST reactor which runs at only 540°C. A US project is the STAR-H2, which will deliver 780°C for hydrogen production and lower temperatures for desalination.

The HTGR is described more fully in section 4.4.

Each 600 MWt module would produce about 200 tonnes of hydrogen per day, which is well matched to the scale of current industrial demand for hydrogen.

The Korean Atomic Energy Research Institute (KAERI) has submitted a Very High Temperature Reactor (VHTR) design to the Generation IV International Forum with a view to hydrogen production from it. This is envisaged as 300 MWt modules, each producing 30,000 tonnes of hydrogen per year. KAERI expects the design concept to be ready in 2008, with operation envisaged in 2020.

KAERI also has a research partnership with China's Tsinghua University focused on hydrogen production, based on China's HTR-10 reactor. A South Korea-USA Nuclear Hydrogen Joint Development Centre involving General Atomics was set up in 2005.

Moving Forward

A 2004 evaluation by JAEA has indicated that by 2010 it expects to confirm the safety of high-temperature reactors and establish operational technology for an IS plant to make hydrogen thermochemically. In April 2004 a coolant outlet temperature of 950°C was achieved in its High-Temperature Engineering Test Reactor (HTTR) – a world first, and opening the way for direct thermochemical hydrogen production.

Meanwhile a pilot plant test project producing hydrogen at 30 m^3/hr from helium heated with 400 kW is planned to test the engineering feasibility of the IS process. By 2015 an IS plant producing 1000 m^3/hr (90 kg/hr, 2 t/day) of hydrogen should be linked to the HTTR to confirm the performance of an integrated production system.

JAEA plans a 600 MW GTHTR300C unit for hydrogen cogeneration using direct cycle gas turbine for electricity and IS process for hydrogen, deploying the first units after 2020. This could produce hydrogen at 60,000 m^3/hr (130 t/day) – "enough for about a million fuel cell vehicles" (at 1 t/day for 7700 cars).

The economics of thermochemical hydrogen production look good. General Atomics projects US$ 1.53/kg, based on a 2400 MWt HTGR operating at 850°C with 42% overall efficiency, and $1.42/kg at 950°C and 52% efficiency (both 10.5% discount rate). At 2003 prices, steam reforming of natural gas yields hydrogen at US$ 1.40/kg, and sequestration of the CO_2 would push this to $1.60/kg. Such a plant could produce 800 t of hydrogen per day, "enough for 1.5 million fuel cell cars" (at 1 t/day for 1800 cars).

In the meantime, hydrogen can be produced by electrolysis of water, using electricity from any source. Non-fossil sources, including

---

[3] Molten fluoride salts are a preferred interface fluid between the nuclear heat source and the chemical plant. The aluminium smelting industry provides substantial experience in managing them safely. The hot molten salt can also be used with secondary helium coolant to generate power via the Brayton cycle, with thermal efficiencies of 48% at 750°C to 59% at 1000°C.

intermittent ones, such as wind and solar, are important possibilities (thereby solving a problem of not being able to store the electricity from those sources). However, the greater efficiency of electrolysis at high temperatures favours a nuclear source for both heat and electricity.

Use of Hydrogen as Fuel

The energy demand for hydrogen production could rival that now used for electricity production.

Burning hydrogen produces only water vapour, with no carbon dioxide or carbon monoxide.

Hydrogen can be burned in a normal internal combustion engine, and some test cars are thus equipped. Trials in aircraft have also been carried out.

However, its main use is likely to be in fuel cells. A fuel cell is conceptually a refuelable battery, making electricity as a direct product of a chemical reaction. But where normal batteries have all the active ingredients built in at the factory, fuel cells are supplied with fuel from an external source.

Fuel cells catalyse the oxidation of hydrogen directly to electricity at relatively low temperatures, and the claimed theoretical efficiency of converting chemical to electrical energy is about 60% (or more). However, in practice about half that has been achieved, except for the higher-temperature solid oxide fuel cells, where it is 46%.

The hydrogen may be stored at very low temperature (cryogenically), at high pressure, or chemically as hydrides. The last is seen to have most potential.

One promising hydride storage system utilizes sodium borohydride ($NaBH_4$) as the energy carrier, with high energy density. The $NaBH_4$ is catalysed to yield its hydrogen, leaving a borate ($NaBO_2$) to be reprocessed.

The first fuel cell electric cars running on hydrogen are expected to be on the fleet market during this decade and the domestic market by 2010. Japan has a goal of 5 million fuel cell vehicles on the road by 2020. (Current electric car technology relies on heavy storage batteries, and the vehicles have limited endurance before slow recharge.)

Current fuel cell design consists of bipolar plates in a frame, and the developer of the proton exchange membrane type, Dr Ballard, suggests that a new geometry is required to bring the cost down and make the technology more widely available to a mass market. Other reviews point out that fuel cells are intrinsically not simple and there are no obvious reasons to expect them to become cheap.

Hydrogen can also be used for stand-alone small-scale stationary generating plants using fuel cells, where higher temperature operation (e.g. of solid oxide fuel cells) and hydrogen storage may be less of a problem, or where it is reticulated like natural gas.

But at present, fuel cells are much more expensive to make than internal combustion engines (burning petrol/gasoline, natural gas or hydrogen).

Other Large-Scale Hydrogen Uses

A peak electricity nuclear system would produce hydrogen at a steady rate and store it underground so that it was used in large banks of fuel cells (e.g. 1000 MWe) at peak demand periods each day. Efficiency would be enhanced if by-product oxygen instead of air were used in the fuel cells.

The initial use of hydrogen for transport is likely to be municipal bus and truck fleets, and prototypes are already on the road in many

Photograph supplied by Ballard Power Systems

Citaro Fuel Cell Bus

parts of the world. These are centrally fuelled, so avoid the need for a retail network, and on-board storage of hydrogen is less of a problem than in cars.

As the scale of hydrogen production increases, more uses in the oil industry become feasible, particularly in the extraction of oil from tar sands. Current practice uses natural gas to produce steam to recover the hydrocarbons, but projected increased production in Canada will exceed available gas supplies. Nuclear-produced hydrogen could thus be used for both heat and hydrogenating the very heavy crude oil.

6.2 NUCLEAR DESALINATION

It is estimated that one fifth of the world's population does not have access to safe drinking water, and that this proportion will increase due to population growth relative to water resources. The worst affected areas are the arid and semi-arid regions of Asia and North Africa. Wars over access to water, not simply energy and mineral resources, are conceivable.

Potable water is in short supply in many parts of the world. Lack of it is set to become a constraint on development in some areas.

Fresh water is a major priority in sustainable development. Where it cannot be obtained from streams and aquifers, desalination of seawater or of mineralized groundwater is required.

Desalination

Most desalination today uses fossil fuels, and thus contributes to increased levels of greenhouse gases. Total world capacity is approaching 30 million m³/day of potable water in some 12,500 plants. Half of these are in the Middle East. The largest produces 450,000 m³/day. Two thirds of the capacity is processing seawater, and one third uses brackish artesian water.

The major technology in use is the multi-stage flash (MSF) distillation process using steam, but reverse osmosis (RO) driven by electric pumps is increasingly significant. MSF gives purer water than RO. A minority of plants use multi-effect distillation (MED) or vapour compression (VC). MSF-RO hybrid plants exploit the best features of each technology for different quality products.

Desalination is energy-intensive. Reverse osmosis needs about 6 kWh of electricity per cubic metre of water, while MSF and MED require heat at 70°C-130°C or

$25 kWh/m^3$ to $200 kWh/m^3$. A variety of low-temperature heat sources may be used, including solar energy. The choice of process generally depends on the relative economic values of fresh water and particular fuels.

Some 10% of Israel's water is desalinated, and one large RO plant provides water at 50 cents US per cubic metre. Malta gets two thirds of its potable water from RO.

Small and medium-sized nuclear reactors are suitable for desalination, often with cogeneration of electricity using low-pressure steam from the turbine and hot sea water feed from the final cooling system. The main opportunities for nuclear plants have been identified as the $80,000 m^3/day$ to $100,000 m^3/day$ and $200,000 m^3/day$ to $500,000 m^3/day$ ranges.

Desalination: Nuclear Experience

Nuclear energy is already being used for desalination, and has the potential for much greater use.

The BN-350 fast reactor at Aktau, Kazakhstan, successfully produced up to 135 MWe of electricity and $80,000 m^3/day$ of potable water over some 27 years, with about 60% of its power being used for heat and desalination. Although the plant was designed as 1000 MWt, it never operated at more than 750 MWt; however, it established the feasibility and reliability of such cogeneration plants. (In fact, oil/gas boilers were used in conjunction with it, and total desalination capacity through ten MED units was $120,000 m^3/day$.)

In Japan, some ten desalination facilities linked to pressurized water reactors operating for electricity production have yielded $1000 m^3/day$ to $3000 m^3/day$ each of potable water, and over 100 reactor-years of experience have accrued. MSF was initially employed, but MED and RO have been found more efficient there. The water is used for the reactors' own cooling systems.

Much relevant experience comes from nuclear plants in Russia, Eastern Europe and Canada where district heating is a by-product.

Large-scale deployment of nuclear desalination on a commercial basis will depend primarily on economic factors. Indicative costs are US$ 0.7-0.9 /m^3$, much the same as fossil-fuelled plants in the same areas. The UN's International Atomic Energy Agency is fostering research and collaboration on the issue, and more than 20 countries are involved.

One obvious strategy is to use all the electricity from power reactors (which tend to run at full capacity) to meet grid load when that is high, and part of it to drive pumps for RO desalination when the grid demand is low.

New Projects

India has been engaged in desalination research since the 1970s and is about to set up a demonstration plant coupled to twin 170 MWe nuclear power reactors (PHWR) at the Madras Atomic Power Station, Kalpakkam, in south-east India. This Nuclear Desalination Demonstration Project will be a hybrid reverse osmosis / multi-stage flash plant, the RO with $1800 m^3/day$ capacity and the higher-quality MSF with $4500 m^3/day$. They will incur a 4 MWe loss in power from the plant. Plants delivering $45,000 m^3/day$ are envisaged, using both kinds of desalination technology.

Russia has embarked on a nuclear desalination project using dual barge-mounted KLT-40 marine reactors (each 150 MWt) and Canadian RO technology to produce potable water.

South Korea has developed a small nuclear reactor design for cogeneration of electricity and potable water at $40,000 m^3/day$. The 330 MWt System-integrated Modular

Advanced Reactor (SMART), an integral PWR, has a long design life and needs refuelling only every three years. The feasibility of building a cogeneration unit employing MSF desalination technology for Madura Island in Indonesia is being studied. Another concept has the SMART reactor coupled to four MED units, each with a thermal-vapour compressor (MED-TVC) and producing a total of 40,000 m^3/day.

Spain, UK, China, India, Pakistan, Egypt, Algeria, Morocco and Tunisia all have projects to build new desalination plants, and the feasibility studies for these often involve nuclear power. Most or all these have requested technical assistance from the IAEA under its technical cooperation project on nuclear power and desalination. A coordinated IAEA research project initiated in 1998 is reviewing reactor designs intended for coupling with desalination systems as well as advanced desalination technologies. Safety and reliability are key requirements. This programme is expected to enable further cost reductions of nuclear desalination.

6.3 NUCLEAR-POWERED SHIPS

Nuclear power is particularly suitable for vessels which need to be at sea for long periods without refuelling, or for powerful submarine propulsion

Work on nuclear marine propulsion started in the 1940s, and the first test reactor started up in the USA in 1953. The first nuclear-powered submarine, USS *Nautilus*, put to sea in 1955, marking the transition of submarines from slow underwater vessels to warships capable of sustaining 20-25 knots, submerged for weeks on end. *Nautilus* led to the parallel development of further *Skate*-class submarines, powered by single pressurized water reactors (PWRs), and an aircraft carrier, USS *Enterprise*, powered by eight PWR units in 1960. A cruiser, USS *Long Beach*, followed in 1961 and was powered by two of these early units. Remarkably, the *Enterprise* remains in service.

By 1962 the US Navy had 26 nuclear submarines operational and 30 under construction. Nuclear power had revolutionized the Navy. The technology was shared with Britain, while French, Russian and Chinese developments proceeded separately.

After the *Skate*-class vessels, reactor development proceeded, and in the USA a single series of standardized designs was built by both Westinghouse and GE, one reactor powering each vessel. Rolls Royce built similar units for Royal Navy submarines and then developed the design further to its PWR-2.

Russia developed both PWR and lead-bismuth cooled reactor designs, the latter not persisting. Eventually four generations of submarine PWRs were utilized, the last entering service in 1995 in the *Severodvinsk* class.

The largest submarines are the 26,500 tonne Russian *Typhoon*-class, powered by twin 190

MWt PWR reactors, though these were superseded by the 24,000 t *Oscar*-II class (e.g. *Kursk*) with the same power plant.

Compared with the excellent safety record of the US nuclear navy, early Soviet endeavours resulted in a number of serious accidents – five where the reactor was irreparably damaged, and more resulting in radiation leaks. However, by the third generation of marine PWRs in the late 1970s, safety had become paramount.

Over 150 ships are powered by more than 220 small nuclear reactors, and more than 12,000 reactor-years of marine operation has been accumulated. Most are submarines, but they range from ice-breakers to aircraft carriers.

Nuclear Naval Fleets

Russia built 248 nuclear submarines and five naval surface vessels powered by 468 reactors between 1950 and 2003. Around 60 of these vessels are still in operation.

At the end of the Cold War, in 1989, there were over 400 nuclear-powered submarines operational or being built. Some 250 of these submarines have now been scrapped and some on order cancelled, due to weapons reduction programmes. Russia and the USA had over 100 each in service, with the UK and France less than 20 each and China 6. The total today is about 160.

The USA has the main navy with nuclear-powered aircraft carriers (11), while both it and Russia have had nuclear-powered cruisers (USA: 9, Russia: 4). The US Navy has accumulated over 5500 reactor-years of accident-free experience, and operates more than 80 nuclear-powered ships (with 105 reactors as of August 2004). Russia has logged 6000 nautical reactor-years.

Civil Vessels

Nuclear propulsion has proven technically and economically essential in the Russian Arctic, where operating conditions are beyond the capability of conventional ice-breakers. The power levels required for breaking ice up to 3 m thick, coupled with refuelling difficulties for other types of vessels, are significant factors. The nuclear fleet has increased Arctic navigation from 2 to 10 months per year, and in the Western Arctic, it is year-round.

The ice-breaker *Lenin* was the world's first nuclear-powered surface vessel (20,000 deadweight tons [dwt]) and remained in service for 30 years, though new reactors were fitted in 1970. It led to a series of larger ice-breakers, the six 23,500 dwt *Arktika*-class, launched from 1975. These powerful vessels have two reactors delivering 56 MW at the propellers and are used in deep Arctic waters. The *Arktika* was the first surface vessel to reach the North Pole, in 1977.

For use in shallow waters such as estuaries and rivers, two shallow-draught *Taymyr*-class ice-breakers of 18,260 dwt with one reactor delivering 38 MW were built in Finland and then fitted with their nuclear steam supply system in Russia. They are built to conform with international safety standards for nuclear vessels and were launched from 1989.

Development of nuclear merchant ships began in the 1950s but on the whole has not been commercially successful. The 22,000 t US-built NS *Savannah*, was commissioned in 1962 and decommissioned eight years later. It was a technical success, but not economically viable. It had a 74 MWt reactor delivering 16.4 MW to the propeller. The German-built 15,000 t *Otto Hahn* cargo ship and research facility sailed some 650,000 nautical miles on 126 voyages in 10 years without any technical problems. It had a 36 MWt reactor delivering 8 MW to the propeller. However, it proved too expensive to operate, and in 1982 it was converted to diesel.

The 8000 t Japanese *Mutsu* was the third civil vessel, put into service in 1970. It had a 36 MWt reactor delivering 8 MW to the propeller. It was dogged by technical and political problems and was an embarrassing failure. These three vessels used reactors with low-enriched uranium fuel (3.7% to 4.4% U-235).

In 1988 the NS *Sevmorput* was commissioned in Russia, mainly to serve northern Siberian ports. It is a 61,900 t lash-carrier (taking lighters to ports with shallow water) and container ship with ice-breaking bow. It is powered by the same KLT-40 reactor as used in larger ice-breakers, delivering 30 propeller MW from the 135 MWt reactor, and it needed refuelling only once to 2003.

Russian experience with nuclear-powered arctic ships totalled 250 reactor-years in 2003. A more powerful ice-breaker of 110 MW net and 55,600 dwt is planned, with further dual-draught ships of 32,400 dwt and 60 MW power.

Power Plants

Naval reactors (with one exception) have been pressurized water types, which differ from commercial reactors producing electricity in that:

- They deliver a lot of power from a very small volume and therefore run on highly-enriched uranium (originally c. 97%, but apparently now 93% in latest US submarines, c. 20% to 25% in some western vessels, and up to 45% in later Russian ones).
- The fuel is not UO_2 but a uranium-zirconium or uranium-aluminium alloy (c. 15% U with 93% enrichment, or more U with less enrichment – e.g. 20% U-235), or a metal-ceramic (e.g. *Kursk*: U-Al zoned 20% to 45% enriched clad in zircaloy, with c. 200 kg U-235 in each 200 MW core).

- They have long core lives, so that refuelling is needed only after 10 or more years, and new cores are designed to last 50 years in carriers and 30-40 years in submarines (e.g. the US *Virginia* class).
- The design enables a compact pressure vessel while maintaining safety.

The long core life is enabled by the relatively high enrichment of the uranium and by incorporating a "burnable poison", such as gadolinium, in the cores, which is progressively depleted as fission products and actinides accumulate, leading to reduced fuel efficiency. The two effects cancel one another out. Long-term integrity of the compact reactor pressure vessel is maintained by providing an internal neutron shield.

The Russian *Alfa*-class submarines had a single liquid metal-cooled reactor (LMR) of 155 MWt and using very highly enriched uranium. These were very fast, but had operational problems in ensuring that the lead-bismuth coolant did not freeze when the reactor was shut down. The design was unsuccessful and used in only eight trouble-plagued vessels.

Reactor power ranges from 10 MWt (in a prototype) up to 200 MWt in the larger submarines and 300 MWt in surface ships such as the *Kirov*-class battle cruisers. The French *Rubis*-class submarines have a 48 MW reactor which needs no refuelling for 30 years. Russia's *Oscar-II* class has two 190 MWt reactors.

The Russian, US and British navies rely on steam turbine propulsion, and the French and Chinese use the turbine to generate electricity for propulsion. Russian ballistic missile submarines as well as all surface ships since the *Enterprise* are powered by two reactors. Other submarines (except some Russian attack subs) are powered by one reactor.

The larger Russian ice-breakers use two KLT-40 nuclear reactors each with 241 or 274 fuel assemblies of 30% to 40% enriched fuel and 3 to 4-year refuelling interval. They drive steam turbines and each produces up to 33 MW (44,000 hp) at the propellers. The large freighter *Sevmorput* uses one of the same units, though it is said to use 90% enriched fuel. For the next generation of Russian ice-breakers, integrated light water reactor designs are being investigated, possibly to replace the conventional PWR.

Decommissioning nuclear-powered submarines has become a major task for US and Russian navies. After defuelling, normal practice is to cut the reactor section from the vessel for disposal in shallow land burial as low-level waste. In Russia the whole vessels, or the sealed reactor sections, sometimes remain stored afloat indefinitely.

Russia is well advanced with plans to build a floating power plant for their far eastern territories. This has 2 x 35 MWe units based on the KLT-40 reactor used in ice-breakers (with refuelling every 4 years).

Future Prospects

In future, constraints on fossil fuel use in transport may bring marine nuclear propulsion into more widespread use.

With increasing attention being given to greenhouse gas emissions arising from burning fossil fuels for international air and marine transport and the excellent safety record of nuclear-powered ships, it is quite conceivable that renewed attention will be given to marine nuclear propulsion.

6.4 SPACE

After a gap of several years, there is a revival of interest in the use of nuclear fission power for space missions.

While Russia has used over 30 fission reactors in space, the USA has flown only one – the SNAP-10A (System for Nuclear Auxiliary Power) in 1965.

From 1959 to 1973 a US nuclear rocket programme – Nuclear Engine for Rocket Vehicle Applications (NERVA) – focused on nuclear power replacing chemical rockets for the latter stages of launches. NERVA used graphite-core reactors to heat hydrogen, which was then expelled through a nozzle. Some 20 engines were tested in Nevada and yielded thrust up to more than half that of the space shuttle launchers. Since then, "nuclear rockets" have been about space propulsion, not launches. The successor to NERVA is today's nuclear thermal rocket (NTR).

Radioisotope Systems

Radioisotope power sources have been used in space since 1961.

Radioisotope thermoelectric generators (**RTGs**) have been the main power source for US space work since 1961. The high decay heat of plutonium-238 (0.56 W/g) enables its use as an electricity source in the RTGs of spacecraft, satellites, navigation beacons and so on. Heat from the oxide fuel is converted to electricity through static thermoelectric elements (solid-state thermocouples), with no moving parts. RTGs are safe, reliable and maintenance-free and can provide heat or electricity for decades under very harsh conditions, particularly where solar power is not feasible.

So far 44 RTGs have powered 24 US space vehicles including Apollo, Pioneer, Viking,

Voyager, Galileo and Ulysses space missions, as well as many civil and military satellites. The Cassini spacecraft carries three RTGs providing 870 watts of power en route to Saturn. Voyager spacecraft, which have sent back pictures of distant planets, have already operated for over 20 years and are expected to send back signals powered by their RTGs for another 15-25 years. The Viking and Rover landers on Mars depended on RTG power sources, as will the Mars Rovers to be launched in 2009.

The latest RTG is a 290-watt system known as the **GPHS RTG**, the thermal power source for this system being the General Purpose Heat Source (GPHS). Each GPHS contains four iridium-clad Pu-238 fuel pellets, stands 5 cm tall, 10 cm square and weighs 1.44 kg. Eighteen GPHS units power one GPHS RTG. The **Multi-Mission RTG** (MMRTG) will use 8 GPHS units producing 2 kW, which can be used to generate 100 watts of electricity.

The **Stirling Radioisotope Generator** (SRG) is based on a 55-watt electric converter powered by one GPHS unit. The hot end of the Stirling converter reaches 650°C, and heated helium drives a free piston reciprocating in a linear alternator, heat being rejected at the cold end of the engine. The AC is then converted to 55 watts DC. This Stirling engine produces about four times as much electric power from the plutonium fuel than an RTG. Thus each SRG will utilize two Stirling converter units with about 500 watts of thermal power supplied by two GPHS units and will deliver 100-120 watts of electric power. The SRG has been extensively tested but has not yet flown in space.

Russia has also developed RTGs using Po-210, and two of these are still in orbit on 1965 Cosmos navigation satellites. But it concentrated on fission reactors for space power systems.

As well as RTGs, Radioactive Heater Units (**RHUs**) are used on satellites and spacecraft to keep instruments warm enough to function efficiently. Their output is only about one watt and they mostly use Pu-238 – typically about 2.7 g of it. Dimensions are about 3 cm long and 2.5 cm diameter, weighing 40 grams. Some 240 have been used so far by the USA, and two are in shut-down Russian Lunar Rovers on the moon. There will be eight on each of the US Mars Rovers launched in 2003.

Both RTGs and RHUs are designed to survive major launch and re-entry accidents intact, as is the SRG.

Fission Systems – Heat

Fission power sources have been used mainly by Russia, but new and more powerful designs are under development in the USA.

For power requirements over 100 kWe, fission systems have a distinct cost advantage over RTGs.

The US SNAP-10A launched in 1965 was a 45 kWt thermal nuclear fission reactor, which produced 650 watts using a thermoelectric converter. It operated for 43 days, after which it had to be shut down due to a satellite (not reactor) malfunction. It remains in orbit.

The last US space reactor initiative was a joint NASA-DOE-Defence Dept. programme developing the SP-100 reactor – a 2 MWt fast reactor unit and thermoelectric system delivering up to 100 kWe as a multi-use power supply for orbiting missions or as a lunar/Martian surface power station. The initiative was terminated in the early 1990s after absorbing nearly $1 billion. The reactor used uranium nitride fuel and was lithium-cooled.

Between 1967 and 1988 the former Soviet Union launched 31 low-powered fission reactors in Radar Ocean Reconnaissance

Satellites (RORSATs) on Cosmos missions. They utilized **thermoelectric** converters to produce electricity, as with the RTGs. Romashka reactors were their initial nuclear power source, a fast spectrum graphite reactor with 90%-enriched uranium carbide fuel operating at high temperature. Later reactors, such as on Cosmos-954, which re-entered over Canada in 1978, had U-Mo fuel rods and a layout similar to the US heatpipe reactors described below.

These were followed by the Topaz reactors with **thermionic** conversion systems, generating about 5 kWe of electricity for on-board uses. This was a US idea developed during the 1960s in Russia.

Topaz-1 was flown in 1987 on two Cosmos missions. It was capable of delivering power for 3-5 years for ocean surveillance. Later Topaz were aiming for 40 kWe via an international project undertaken largely in the USA from 1990. Two Topaz-2 reactors (without fuel) were sold to the USA in 1992. Budget restrictions in 1993 forced cancellation of a Nuclear Electric Propulsion Spaceflight Test Programme associated with this.

Fission Systems – Propulsion

For spacecraft propulsion, once launched, some experience has been gained with **nuclear thermal** propulsion systems (NTR). Nuclear fission heats a hydrogen propellant, which is stored as liquid in cooled tanks. The hot gas (about 2500°C) is expelled through a nozzle to give thrust (which may be augmented by injection of liquid oxygen into the supersonic hydrogen exhaust). This is more efficient than chemical reactions. Bimodal versions will run electrical systems on board a spacecraft, including powerful radars, as well as providing propulsion. Compared with nuclear electric plasma systems, these have much more thrust for shorter periods and can be used for launches and landings.

However, attention is now turning to **nuclear electric** systems, where nuclear reactors are a heat source for electric ion drives expelling plasma out of a nozzle to propel spacecraft already in space. Superconducting magnetic cells ionize hydrogen or xenon, heat it to extremely high temperatures (millions °C), accelerate it and expel it at very high velocity (e.g. 30 km/sec) to provide thrust.

Research for one version, the Variable Specific Impulse Magnetoplasma Rocket (VASIMR), draws on that for magnetically-confined fusion power (tokamak) for electricity generation, but here the plasma is deliberately leaked to give thrust. The system works most efficiently at low thrust (which can be sustained), with small plasma flow, but high thrust operation is possible. It is very efficient, with 99% conversion of electric to kinetic energy.

Heatpipe Power System (HPS) reactors are compact fast reactors producing up to 100 kWe for about ten years to power a spacecraft or planetary surface vehicle. They have been developed since 1994 at the Los Alamos National Laboratory as a robust and low technical risk system with an emphasis on high reliability and safety. They employ heatpipes to transfer energy from the reactor core to make electricity using Stirling or Brayton cycle converters.

Energy from fission is conducted from the fuel pins to the heatpipes filled with sodium vapour which carry it to the heat exchangers and thence in hot gas to the power conversion systems to make electricity. The gas is 72% helium and 28% xenon.

The **SAFE-400** space fission reactor (Safe Affordable Fission Engine) is a 400 kWt HPS producing 100 kWe to power a space vehicle using two Brayton power systems – gas turbines driven directly by the hot gas from the reactor. SAFE has also been tested with an electric ion drive.

A smaller version of this kind of reactor is the **HOMER-15** – the Heatpipe-Operated Mars Exploration Reactor. It is a 15 kW thermal unit similar to the larger SAFE model, it stands 2.4 m tall including its heat exchanger and has a 3 kWe Stirling engine (see above). Total mass of the reactor system is 214 kg, and the diameter is 41 cm.

Project Prometheus 2003

In 2002 NASA launched Project Prometheus, whose purpose is to enable a major step change in the capability of space missions. Nuclear-powered space travel will be much faster than is now possible, and this would enable manned missions to Mars.

One part of Prometheus, which is a NASA project with substantial involvement by DOE in the nuclear area, is to develop the Multi-Mission Thermoelectric Generator and the Stirling Radioisotope Generator described in the RTG section above.

A more radical objective of Prometheus is to produce a space fission reactor system, such as those described above for both power and propulsion that is safe to launch and which will operate for many years. This will have much greater power than RTGs. Power of 100 kW is envisaged for a nuclear electric propulsion system driven by plasma.

See also: WNA information paper Nuclear Reactors for Space

6.5 RESEARCH REACTORS FOR MAKING RADIOISOTOPES

Many of the world's nuclear reactors are used for research and training, materials testing, or the production of radioisotopes for medicine and industry. These are much smaller than power reactors or those propelling ships, and many are on university campuses.

Research reactors comprise a wide range of civil and commercial nuclear reactors, which are generally not used for power generation. The primary purpose of research reactors is to provide a neutron source for research and other purposes. Their output (neutron beams) can have different characteristics depending on use. They are small relative to power reactors, whose primary function is to produce heat to make electricity. Their power is designated in megawatts (or kilowatts) thermal (MWt), but here we will use simply MW (or kW). Most range up to 100 MW, compared with 3000 MW (i.e. 1000 MWe) for a typical power reactor. In fact the total power of the world's 283 research reactors is little over 3000 MW.

Research reactors are simpler than power reactors and operate at lower temperatures. They need far less fuel, and far fewer fission products build up as the fuel is used. On the other hand, their fuel requires more highly enriched uranium, typically up to 20% U-235, although many older ones used 93% U-235. They also have a very high power density in the core, which requires special design features. Like power reactors, the core needs cooling, and usually a moderator is required to slow down the neutrons and enhance fission. As neutron production is their main function, most research reactors also need a reflector to reduce neutron loss from the core.

There are about 280 such reactors operating in 56 countries.

Types of Research Reactors

There is a much wider array of designs in use for research reactors than for power reactors, where 80% of the world's plants are of just two similar types. They also have different operating modes, producing energy which may be steady or pulsed.

A common design (over 65 units) is the pool type reactor, where the core is a cluster of fuel elements sitting in a large pool of water. Among the fuel elements are control rods and empty channels for experimental materials. Each element comprises several (e.g. 18) curved aluminium-clad fuel plates in a vertical box. The water both moderates and cools the reactor, and graphite or beryllium is generally used for the reflector, although other materials may be used. Apertures to access the neutron beams are set in the wall of the pool. Tank type research reactors (32 units) are similar, except that cooling is more active.

The TRIGA reactor is another common design (40 units). The core consists of 60-100 cylindrical fuel elements about 36 mm diameter with aluminium cladding enclosing a mixture of uranium fuel and zirconium hydride (as moderator). It sits in a pool of water and generally uses graphite or beryllium as a reflector. This kind of reactor can safely be pulsed to very high power levels (e.g. 25,000 MW) for fractions of a second. Its fuel gives the TRIGA a very strong negative temperature coefficient, and the rapid increase in power is quickly cut short by a negative reactivity effect of the hydride moderator.

Other designs are moderated by heavy water (12 units) or graphite. A few are fast reactors, which require no moderator and can use a mixture of uranium and plutonium as fuel. Homogenous-type reactors have a core comprising a solution of uranium salts as a liquid, contained in a tank about 300 mm diameter. The simple design made them popular early on, but only five are now operating.

Research reactors have a wide range of uses, including analysis and testing of materials, and production of radioisotopes. Their capabilities are applied in many fields, within the nuclear industry as well as in fusion research, environmental science, advanced materials development, drug design and nuclear medicine.

The IAEA lists several categories of broadly-classified research reactors. They include critical assemblies – usually zero power (60), test reactors (23), training facilities (37), two prototypes and even one producing electricity. But most (160) are largely for research, although some may also produce radioisotopes. As expensive scientific facilities, they tend to be multi-purpose, and many have been operating for more than 30 years.

Russia has most research reactors (62), followed by the USA (54), Japan (18), France (15), Germany (14) and China (13). Many small and developing countries also have research reactors, including Bangladesh, Algeria, Colombia, Ghana, Nigeria, Jamaica, Libya, Thailand and Vietnam. About 20 more reactors are planned or under construction, and 361 have been shut down or decommissioned, about half of these in the USA. Many research reactors were built in the 1960s and 1970s. The peak number operating was in 1975, with 373 in 55 countries.

Uses

Neutron beams are uniquely suited to studying the structure and dynamics of materials at the atomic level. For example, "neutron scattering" is used to examine samples under different conditions, such as variations in vacuum pressure, high temperature, low temperature and magnetic field, essentially under real-world conditions.

Using neutron activation analysis, it is possible to determine accurately the composition of minute quantities of material. Atoms in a sample are made radioactive by exposure to neutrons in a reactor. The characteristic radiation each element emits can then be detected.

Neutron activation is also used to produce the radioisotopes, widely used in industry and medicine, by bombarding particular elements with neutrons. For example, yttrium-90 microspheres to treat liver cancer are produced by bombarding yttrium-89 with neutrons. The most widely used isotope in nuclear medicine is technetium-99, a decay product of molybdenum-99. It is produced by irradiating U-235 foil with neutrons and then separating the molybdenum from the other fission products in a hot cell.

Research reactors can also be used for industrial processing. Neutron transmutation doping makes silicon crystals more electrically conductive for use in electronic components. In test reactors, materials are subject to intense neutron irradiation to study changes. For instance, some steels become brittle, and alloys which resist embrittlement must be used in nuclear reactors.

Like power reactors, research reactors are covered by IAEA safety inspections and safeguards, because of their potential for making nuclear weapons. India's 1974 explosion was the result of plutonium production in a large, but internationally unsupervised, research reactor.

Fuels

Fuel assemblies are typically plates or cylinders of uranium-aluminium alloy (U-Al) clad with pure aluminium. They are different from the ceramic UO_2 pellets enclosed in Zircaloy cladding type assemblies often used in power reactors. Only a few kilograms of uranium are needed to fuel a research reactor, albeit more

highly enriched, compared with perhaps a hundred tonnes in a power reactor.

Some operate with high-enriched uranium fuel, and international efforts are under way to substitute low-enriched fuel.

Highly-enriched uranium (HEU – >20% U-235) allows more compact cores, with high neutron fluxes and also longer times between refuelling. Therefore many reactors up to the 1970s used it, and in 2004 more than 60 civilian research reactors still did so.

Since the early 1970s security concerns have grown, especially since many research reactors are located at universities and other civilian locations with much lower security than military weapons establishments where much larger quantities of HEU exist. Since 1978 only one reactor, the FRM-II at Garching, Germany, has been built with HEU fuel, while 21 have been commissioned on low-enriched (LEU) fuel in 16 countries.

The question of enrichment was a major focus of the UN-sponsored International Nuclear Fuel Cycle Evaluation in 1980. It concluded that to guard against weapons proliferation from the HEU fuels then commonly used in research reactors, enrichment should be reduced to no more than 20% U-235. This followed a similar initiative by the USA in 1978 when its programme for Reduced Enrichment for Research and Test Reactors (RERTR) was launched.

Most research reactors using HEU fuel were supplied by the USA and Russia, hence efforts to deal with the problem are largely their initiative. The RERTR programme concentrates on reactors over 1 MW which have significant fuel requirements.

These programmes have led to the development and qualification of new high-density LEU fuels. The original fuel density was about 1.3 g/cm³ to 1.7 g/cm³ uranium. Lowering the enrichment meant that the density had to be increased. Initially this was to 2.3 g/cm³ to 3.2 g/cm³ with existing U-Al fuel types.

To late 2004, 38 research reactors (11 in the USA) either have been or are being converted to low-enriched uranium silicide fuel, and another 36 are convertible using present fuels. Thirty-one more, mostly Russian designs, need higher-density fuels not yet available. The goal is to convert 105 reactors by 2013. Some other HEU reactors are expected to close down by then. US exports of HEU declined from 700 kg/yr in the mid-1970s to almost zero by 1993.

The Soviet Union made similar efforts from 1978, and produced fuel of 2.5 g/cm³ with enrichment reduced from 90% to 36%. It largely stopped exports of 90% enriched fuel in the 1980s. However, no Russian research reactor has yet been converted to LEU.

The first generation of new LEU fuels used uranium and silicon (U_3Si_2-Al – uranium silicide dispersed in aluminium) at 4.8 g/cm³. There have been successful tests with denser U_3Si-Al fuel plates up to 6.1 g/cm³, but US development of these silicide fuels ceased in 1989 and did not recommence until 1996.

An international effort is under way to develop, qualify and license a high density fuel based on U-Mo alloy dispersed in aluminium, with a density of 6 g/cm³ to 8 g/cm³. The principal organizations involved are the US RERTR programme at Argonne National Laboratory (ANL) since 1996, the French U-Mo Group (CEA, CERCA, COGEMA, Framatome-ANP and Technicatome) since 1999 and the Argentine Atomic Energy Commission (CNEA) since 2000. This development work has been undertaken to provide fuels which can extend

the use of LEU to those reactors requiring higher densities than available in silicide dispersions and to provide a fuel that can be more easily reprocessed than the silicide type. Approval of this fuel was expected in 2006, but tests since 2003 have failed to confirm performance due to unstable swelling under high irradiation, and the target is now 2010.

In Russia, a parallel RERTR programme funded by the Russian Ministry of Atomic Energy (MINATOM) and the US RERTR programme has been working since 1999 to develop U-Mo dispersion fuel with a density of 2 g/cm³ to 6 g/cm³ for use in Russian-designed research and test reactors. However, this too has not fulfilled expectations.

In a further stage of U-Mo fuel development, which has become the main priority, ANL, CEA and CNEA are testing U-Mo fuel in a monolithic form, instead of a dispersion of U-Mo in aluminium. The uranium density is 15.6 g/cm³, and this would enable every research reactor in the world to convert from HEU to LEU fuel without loss of performance. The target date for availability is 2010.

All fuel is aluminium-clad.

Spent Fuel

U-Al fuels can be reprocessed by Cogema in France, and U-Mo fuels may also be reprocessed there. U-Si and TRIGA fuels are not readily reprocessed in conventional facilities. However, at least one commercial operator has confirmed that U-Si fuels may be reprocessed in existing plants if diluted with appropriate quantities of other fuels, such as U-Al.

To answer concerns about interim storage of spent research fuel around the world, the USA launched a programme to take back US-origin spent fuel for disposal, and nearly half a tonne of U-235 from such HEU fuel has been

OTHER NUCLEAR ENERGY APPLICATIONS

109

returned. By the time the programme was due to end with fuel discharged in 2006, U-Mo fuel was expected to be available. Due to the slippage in target date, the US take-back programme has now been extended by 10 years.

Disposal of high-enriched or even 20% enriched fuel needs to address problems of criticality and requires the use of neutron absorbers or diluting or spreading it out in some way.

In Russia, a parallel trilateral programme involving IAEA and the USA is intended to move 2 t of HEU and 2.5 t of LEU spent fuel to the Mayak reprocessing complex near Chelyabinsk over the ten years to 2012. This Russian Research Reactor Fuel Return Programme (RRR FRT) envisages 38 shipments (of both fresh and spent fuel) from 10 countries from 2005 to 2008, then at least 8 shipments from six countries to remove all HEU fuel discharged before reactors convert to LEU or shut down. Seventeen countries have Soviet-supplied research reactors, and there are 25 such reactors outside Russia, 15 of them still operational. Since Libya joined the programme in 2004, only North Korea objects to it.

ENVIRONMENT, HEALTH AND SAFETY ISSUES

Environmental and health consequences of electricity generation are external costs – those which are quantifiable but do not appear in the utility's accounts. Hence they are not passed on to the consumer, but are borne by society at large. They include particularly the effects of air pollution on human health, crop yields and buildings, as well as occupational disease and accidents. Though they are even harder to quantify and evaluate than the others, external costs include effects on ecosystems and the impact of global warming.

The need for clean electricity generation has never been more evident, nor popularly supported.

Production of electricity from any form of primary energy has some environmental effect. A balanced assessment of nuclear power requires comparison of its environmental effects with those of the principal alternative, coal-fired electricity generation, as well as with other options.

7.1 GREENHOUSE GAS EMISSIONS

Greenhouse here refers to the effect of certain trace gases in the Earth's atmosphere so that long-wave radiation, such as heat from the Earth's surface, is trapped. A build-up of greenhouse gases, notably CO_2, appears to be causing a warming of the climate in many parts of the world, which if continued will cause changes in weather patterns and other profound changes. Much of the greenhouse effect is due to carbon dioxide[1].

While our understanding of relevant processes is advancing, we do not know how much carbon dioxide the environment can absorb, nor how long-term global CO_2 balance is maintained. However, scientists are increasingly concerned about the steady worldwide build-up of CO_2 levels in the atmosphere, and political initiatives reflect this. The build-up is occurring as the world's carbon-based fossil fuels are being burned and rapidly converted to atmospheric CO_2, for example, in motor vehicles, domestic and industrial furnaces, and electric power generation.

Figure 20: Greenhouse gas emissions for electricity production

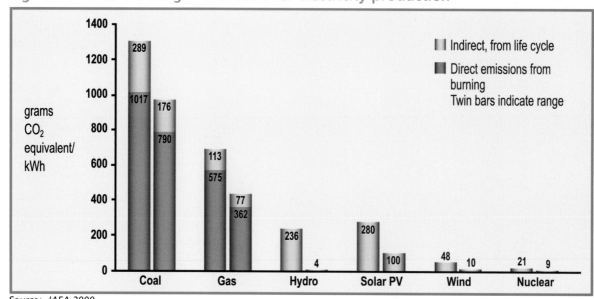

Source: IAEA 2000.

[1] CO_2 constitutes only 0.035% (380 ppm) of the atmosphere. An increase from 280 to 380 ppm has already occurred since the beginning of the Industrial Revolution.

Progressive clearing of the world's forests also contributes to the greenhouse effect by diminishing the removal of atmospheric CO_2 by photosynthesis.

As early as 1977 a USA National Academy of Sciences report concluded that "the primary limiting factor on energy production from fossil fuels over the next few centuries may turn out to be the climatic effects of the release of carbon dioxide". Today this is conventional wisdom. **The inexorable increase of CO_2 levels in the atmosphere, coupled with concern about their possible climate effect, is now a very significant factor in the comparison of coal and nuclear power for producing electricity.**

Worldwide emissions of CO_2 from burning fossil fuels total about 25 billion tonnes per year. About 38% of this is from coal and about 43% from oil. Every 1000 MWe power station running on black coal produces CO_2 emissions of about 7 million tonnes per year. If brown coal is used, the amount is about 9 million tonnes. Nuclear fission does not produce CO_2, while emissions from other parts of the fuel cycle (e.g. uranium mining and enrichment) amount to about 2% of those from using coal, and some audited figures show considerably less than this.

Every 22 tonnes of uranium (26 t U_3O_8) used[2] saves about 1 million tonnes of CO_2 relative to coal.

There is now widespread agreement that we need resource strategies which will minimize CO_2 build-up. With respect to base-load electricity generation, increased use of uranium as a fuel is the most obvious such strategy, utilizing proven technology on the scale required (see also Figure 20).

---

[2] in a light water reactor

7.2 OTHER ENVIRONMENTAL EFFECTS OF ELECTRICITY GENERATION

At a uranium mine ordinary operating procedures normally ensure that there is no significant water or air pollution. The environmental effect of coal mining today is also small, except that more extensive areas may require subsequent rehabilitation, and in certain areas acid mine drainage due to oxidation of sulphur can be a problem. The effects of uranium mining are discussed more fully in section 4.1.

Small amounts of **radioactivity** are released to the atmosphere from both coal-fired and nuclear power stations. In the case of coal combustion small quantities of uranium, radium and thorium present in the coal cause the fly ash to be radioactive, the level varying considerably. Nuclear power stations and reprocessing plants release small quantities of radioactive gases (e.g. krypton-85 and xenon-133) and iodine-131, which may be detectable in the environment with sophisticated monitoring and analytical equipment. Steps are being taken to reduce further emissions of both fly ash from coal-fired power stations and radionuclides from nuclear power stations and other plants. At present neither constitutes a significant environmental problem.

As outlined in sections 5.3-5.5, solid high-level waste from nuclear power stations is stored for 40-50 years while the radioactivity decays to less than 1% of its original level. Then it will be finally disposed of deep underground and well away from the biosphere. There has been no pollution from such material and nor is any likely, either short- or very long-term.

Intermediate-level waste is placed in underground repositories. Low-level waste is generally buried more conventionally.

Radioactive fly ash from coal-fired power stations has in the past had a much greater environmental impact, largely because it was not perceived as a problem and appropriate action was not taken. Today most fly ash is removed from stack gases and buried where seepage and run-off can be controlled.

Waste heat produced due to the intrinsic inefficiency of energy conversion, and hence as a by-product of power generation, is much the same whether coal or uranium is the primary fuel. The thermal efficiency of coal-fired power stations ranges from about 20% to a possible 40%, with newer ones typically giving better than 33%. That of nuclear stations mostly ranges from 29% to 38% with the common light water reactor today giving about 34%. There is no reason for preferring one fuel over the other on account of waste heat. This is the case whether power station cooling is by water from a stream or estuary, or using atmospheric cooling towers. In any case this heat need not always be "waste". In colder climates, district heating and agricultural uses are increasingly found. These decrease the extent to which local fogs result from the release of heat to the environment.

The main environmental matter relevant to power generation is the production of carbon dioxide (CO_2) and sulphur dioxide (SO_2) as a result of coal-fired electricity generation. When coal of say 2.5% sulphur is used to produce the electricity for one person in an industrialized country for one year, then about 9 t of CO_2 and 120 kg of SO_2 are produced (see Figure 6).

Sulphur dioxide emissions arise from the combustion of fossil fuels containing sulphur, as many do. Released in large quantities to the atmosphere, it can cause (sulphuric) "acid rains" in areas downwind. In the northern hemisphere many millions of tonnes of SO_2 are released annually from electricity generation, though such pollution has been dramatically reduced. The acid rain (rainwater having a pH of 4 and lower) in north-eastern USA and Scandinavia causes ecological changes and economic loss. In the UK and the USA, electric power utilities at first sought to minimize this by increasing their use of natural gas, but costs now work against this.

It is possible to remove a lot of the SO_2 from coal stack gases using flue gas desulphurization equipment, but the cost is considerable. Power utilities have spent many billions of dollars on this. On the other hand, between 1980 and 1986 SO_2 emissions in France were halved simply by replacing fossil fuel power stations with nuclear ones. At the same time, electricity production increased 40% and France became a significant exporter of electricity.

Oxides of nitrogen (NO_X) from fossil fuel power stations operating at high temperatures are also an environmental problem. If high levels of hydrocarbons are present in the air, nitrogen oxides react with these to form photochemical smog. Moreover, oxides of nitrogen have an adverse effect on the Earth's ozone layer, increasing the amount of ultraviolet light reaching the Earth's surface.

7.3 HEALTH AND ENVIRONMENTAL EFFECTS OF POWER GENERATION

Here the emphasis is on comparing nuclear power with coal-fired power plants for electricity. Both occupational and environmental health effects are considered along with risks.

Traditionally, occupational health risks have been measured in terms of immediate accident, especially fatality rates. However, today, and particularly in relation to nuclear power, there is an increased emphasis on less obvious or delayed effects of exposure to cancer-inducing substances and radiation.

Many occupational accident statistics have been generated over the last 40 years of nuclear reactor operations in the USA and the UK. These can be compared with those from coal-fired electricity generation. All show that nuclear power is distinctly the safer means of electric power generation in this respect. Two simple sets of figures are quoted in Tables 12 & 12A. A major reason for coal showing up unfavourably is the huge amount of it which must be mined and transported to supply even a single large power station. Mining and multiple handling of so much material of any kind involves hazards, and these are reflected in the statistics.

Health risks in uranium mining are largely discussed in section 4.1. In the 1950s exposure of miners to radon gas led to a higher incidence of lung cancer. For over 40 years, however, exposure to high levels of radon has not been a feature of uranium or other mines. Today the presence of some radon around a uranium mining operation and some dust bearing radioactive decay products as well as the hazards of inhaled coal dust in a coal mine are well understood. In both cases, using the best current practice, the health hazards to miners are very small and certainly less than the risks of industrial accidents.

In other parts of the nuclear fuel cycle, radiation hazards to workers are low, and industrial accidents are few. Certainly nuclear power generation is not completely free of hazards in the occupational sense, but it does appear to be far safer than other forms of energy conversion. Table 12 covers more than 20 years.

The occurrence of cancer is not uniform across the world population, and because of local differences it is not easy to see whether there is any association between low occupational radiation doses and possible excess cancers. However, this question has been studied closely in a number of areas and work is continuing. So far no conclusive evidence has emerged to indicate that cancers are more

Table 12: Comparison of accident statistics in primary energy production.
(Electricity generation accounts for about 40% of total primary energy).

Fuel	Immediate fatalities 1970-1992	Who?	Normalized to deaths per TWy* electricity
Coal	6400	workers	342
Natural gas	1200	workers & public	85
Hydro	4000	public	883
Nuclear	31	workers	8

Basis: per million MWe operating for one year (i.e. about three times world nuclear power capacity), not including plant construction, based on historic data – which is unlikely to represent current safety levels in any of the industries concerned. The data in this column was published in 2001 but is consistent with that from 1996-1997, where it is pointed out that the coal total would be about ten times greater if accidents with less than five fatalities were included.

Source: Ball, Roberts & Simpson, Research Report #20, Centre for Environmental & Risk Management, University of East Anglia, 1994; Hirschberg et al, Paul Scherrer Institut, 1996; in: IAEA, Sustainable Development and Nuclear Power, 1997; Severe Accidents in the Energy Sector, Paul Scherrer Institut, 2001.

Table 12A: The hazards of using energy: some energy-related accidents since 1977

PLACE	YEAR	NUMBER KILLED	COMMENTS
Machhu II, India	1979	2500	hydroelectric dam failure
Hirakud, India	1980	1000	hydroelectric dam failure
Ortuella, Spain	1980	70	gas explosion
Donbass, Ukraine	1980	68	coal mine methane explosion
Israel	1982	89	gas explosion
Guavio, Colombia	1983	160	hydroelectric dam failure
Nile R, Egypt	1983	317	LPG explosion
Cubatao, Brazil	1984	508	oil fire
Mexico City	1984	498	LPG explosion
Tbilisi, Russia	1984	100	gas explosion
Northern Taiwan	1984	314	3 coal mine accidents
Chernobyl, Ukraine	1986	56+	nuclear reactor accident
Piper Alpha, North Sea	1988	167	explosion of offshore oil platform
Asha-ufa, Siberia	1989	600	LPG pipeline leak and fire
Dobrnja, Yugoslavia	1990	178	coal mine
Hongton, Shaanxi, China	1991	147	coal mine methane explosion
Belci, Romania	1991	116	hydroelectric dam failure
Kozlu, Turkey	1992	272	coal mine methane explosion
Cuenca, Ecuador	1993	200	coal mine
Durunkha, Egypt	1994	580	fuel depot hit by lightning
Seoul, S.Korea	1994	500	oil fire
Minanao, Philippines	1994	90	coal mine
Dhanbad, India	1995	70	coal mine
Taegu, S.Korea	1995	100	oil & gas explosion
Spitsbergen, Russia	1996	141	coal mine
Henan, China	1996	84	coal mine methane explosion
Datong, China	1996	114	coal mine methane explosion
Henan, China	1997	89	coal mine methane explosion
Fushun, China	1997	68	coal mine methane explosion
Kuzbass, Siberia	1997	67	coal mine methane explosion
Huainan, China	1997	89	coal mine methane explosion
Donbass, Ukraine	1998	63	coal mine methane explosion
Liaoning, China	1998	71	coal mine methane explosion
Warri, Nigeria	1998	500+	oil pipeline leak and fire
Donbass, Ukraine	1999	50+	coal mine methane explosion
Donbass, Ukraine	2000	80	coal mine methane explosion
Muchonggou, Guizhou, China	2000	162	coal mine methane explosion
Jixi, China	2002	124	coal mine methane explosion
Gaoqiao, SW China	2003	234	gas well blowout with H_2S
Kuzbass, Russia	2004	47	coal mine methane explosion
Donbass, Ukraine	2004	36	coal mine methane explosion
Henan, China	2004	148	coal mine methane explosion
Chenjiashan, Shaanxi, China	2004	166	coal mine methane explosion
Sunjiawan, Liaoning, China	2005	215	coal mine methane explosion
Fukang, Xinjiang, China	2005	83	coal mine methane explosion
Xingning, Guangdong, China	2005	102	coal mine flooding
Dongfeng, Heilongjiang, China	2005	164	coal mine methane explosion

LPG and oil accidents with less than 300 fatalities, and coal mine accidents with less than 100 fatalities are generally not shown unless recent. Coal mining deaths range from 0.009 per million tonnes of coal mined in Australia, through 0.034 in the USA, to 4 in China, and 7 in Ukraine. China's total death toll from coal mining averages well over 5000 per year – official figures give 5300 in 2000, 5670 in 2001, 7200 in 2003 and 6027 in 2004. Ukraine's coal mine death toll is over 200 per year (e.g. 1999: 274, 1998: 360, 1995: 339, 1992: 459).

Sources: contemporary media reports, Paul Scherrer Inst, 1998 report.

frequent in radiation workers than in other people of similar ages in western countries. Nor, incidentally, are they greater in people exposed to very high natural levels of radiation in certain parts of the world – significantly higher than levels allowed in industry. At the low levels of exposure and dose rates involved in the nuclear industry, the effects are probabilistic rather than measurable, as described in section 7.4.

Environmental (non-occupational) health effects are qualitatively similar to those affecting workers in the industry. Popular concern about ionizing radiation initially grew out of the testing of nuclear weapons. Correspondingly, these tests provided the nuclear power industry with a strong awareness of radiation hazards. Fortunately radioactivity is readily measurable and its effects fairly well understood compared with those of other hazards with delayed effects – including virtually all chemical cancer-inducing substances. Radiation is a weak carcinogen.

The contrast between air quality effects from coal burning for electricity and increased radiation from nuclear power is very marked: a person living next to a nuclear power plant receives less radiation from it than from a few hours flying each year (see Table 13). On the other hand, anyone downwind of a coal-fired power plant can expect it to have an effect on the air quality.

7.4 RADIATION

Table 13 shows some typical levels and sources of radiation exposure. The contribution from the ground and buildings varies from place to place. In most parts of the world levels range up to 3 mSv/yr. Citizens of Cornwall, UK, receive an average of about 7 mSv/yr. Hundreds of thousands of people in India, Brazil and Sudan receive up to 40 mSv/yr. Several places are known in Iran, India and Europe where natural background radiation gives an annual dose of more than 50 mSv, and in Ramsar, Iran, it can give up to 260 mSv. Lifetime doses from natural radiation range up to several thousand millisieverts. However, there is no evidence of increased cancers or other health problems arising from these high natural levels.

Cosmic radiation dose varies with altitude and latitude. Aircrew can receive up to about 5 mSv/yr from their hours in the air, and frequent flyers can score a similar increment. In contrast, UK citizens receive about 0.0003 mSv/yr from nuclear power generation. Appendix 1 gives further background to the topic of radiation and its measurement.

In practice, radiation protection is based on the understanding that small increases over natural levels of exposure are not likely to be harmful but should be kept to a minimum. To put this into practice the International Commission for Radiological Protection (ICRP) has established recommended standards of protection based on three basic principles:

- **Justification**. No practice involving exposure to radiation should be adopted unless it produces a net benefit to those exposed or to society generally.

- **Optimization**. Radiation doses and risks should be kept as low as reasonably achievable (ALARA), economic and social factors being taken into account.

- **Limitation**. The exposure of individuals

should be subject to dose or risk limits above which the radiation risk would be deemed unacceptable.

These principles apply to the potential for accidental exposures as well as predictable normal exposures.

Underlying these principles is the application of the "linear hypothesis", based on the idea that any level of radiation dose, no matter how low, involves the possibility of risk to human health.

This assumption enables "risk factors" derived from studies of high radiation dose to populations (e.g. from Japanese survivors of atomic bombs) to be used in determining the risk to an individual from low doses[3]. However the weight of scientific evidence does not indicate any cancer risk or immediate effects at doses below 50 mSv in a short time or at about 100 mSv per year. At lower doses and dose rates (up to at least 10 mSv/yr) the evidence suggests that beneficial effects are at least as likely as harmful ones.

Table 13: Ionizing radiation

The Earth is radioactive, due to the decay of natural long-lived radioisotopes. Radioactive decay results in the release of ionizing radiation. As well as the Earth's radioactivity we are naturally subject to cosmic radiation from space. In addition to both these, we collect some radiation doses from artificial sources such as X-rays. We may also collect an increased cosmic radiation dose by participating in high altitude activities such as flying or skiing. The average adult contains about 13 mg of radioactive potassium-40 in body tissue – we therefore even irradiate one another at close quarters!

The relative importance of these various sources is indicated in the Table below. Types of radiation and units for measuring it are outlined in Appendix 1.

	Typical μSv/yr	Range
Natural:		
Terrestrial + house: radon	200	200-100,000
Terrestrial + house: gamma	600	100-1000
Cosmic (at sea level)	300	
+20 for every 100m elevation		0-500
Food, drink & body tissue	400	100-1000
Total	1500 (plus altitude adjustment)	
Artificial:		
From nuclear weapons tests	3	
Medical (X-ray, CT etc. average)	370	up to 75,000
From nuclear energy	0.3	
From coal burning	0.1	
From household appliances	0.4	
Total	375	
Behavioural:		
Skiing holiday	8/wk	
Air travel in jet airliner	1.5-5/hr	up to 5000/yr

The International Commission for Radiological Protection recommends, in addition to background, the following exposure limits:

| For general public | 1000 (i.e. 1 mSv/yr) |
| For nuclear worker | 20,000 (i.e. 20 mSv/yr) averaged over 5 consecutive years |

Sources: Australian Radiation Protection & Nuclear Safety Agency, National Radiation Protection Board (UK), Australian Nuclear Science & Technology Organization, various.

[3] ICRP Publication 60.

Based on the three conservative principles, ICRP recommends that the additional dose above natural background and excluding medical exposure should be limited to prescribed levels. These are: 1 mSv/yr for members of the public, and 20 mSv/yr averaged over 5 years for radiation workers who are required to work under closely-monitored conditions (see Table 13).

The actual **level of individual risk** at the ICRP recommended limit for general public exposure is very small (it is calculated to result in about 1 fatal cancer per year in a population of 20,000 people) and impossible to confirm directly. In the Chernobyl accident (see section 7.5), a large number of people were subject to significantly increased radiation exposure, the actual doses being approximately known. In due course this tragedy may result in a better understanding of the effects, if any, of exposure to various levels of radiation. At present much of our knowledge about the effect of radiation on people is derived from the survivors of the Hiroshima and Nagasaki bombings in 1945, where the doses received were very brief and also difficult to estimate. Certainly there was a clear increase in certain types of leukaemia and lymphoma and of solid cancers among the survivors. Progressively there is more information based on exposure with low dose rate, where the body has time to repair damage.

The body has defence mechanisms against damage induced by radiation as well as by chemical carcinogens. These can be stimulated by low levels of exposure, or overwhelmed by very high levels[4].

Plutonium is sometimes seen as a particular concern. It is separated from spent fuel by reprocessing, as discussed in section 5.2. Plutonium has been called the most toxic element known to man and therefore represented as a hazard that we should do without. However it is pertinent to compare its toxicity with that of other materials with which we live. If swallowed, plutonium is much less toxic than cyanide or lead arsenate and about twice as toxic as the concentrate of caffeine from coffee. Its main danger comes if inhaled as a fine dust and absorbed through the lungs. This would increase the likelihood of cancer 15 or more years afterwards. However, as a counterpoint to the folklore about plutonium is the fact that about seven tonnes of it were dispersed in the upper atmosphere by nuclear weapons testing over the 30 years following World War II without identifiable ill effects.

The health effects of exposure both to radiation and to chemical cancer-inducing agents or toxins must be considered in relation to time. We should be concerned not only about the effects on people presently living, but also about the cumulative effects of actions today over many generations. Some radioactive materials which reach the environment decay to safe levels within days, weeks or a few years, while others continue their effect for a long time, as do some chemical cancer-inducing agents and toxins. Certainly this is true of the chemical toxicity of heavy metals such as mercury, cadmium and lead, these of course being a natural part of the human environment anyway, like radiation, but maintaining their toxicity forever. The essential task for those in government and industry is to prevent excessive amounts of such toxins harming people, now or in the future. Standards are set in the light of research on environmental pathways by which people might ultimately be affected.

About 60 years ago it was discovered that ionizing radiation could induce **genetic**

---

[4] Tens of thousands of people in each technically-advanced country work in medical and industrial environments where they may be exposed to radiation above background levels. Accordingly they wear monitoring "badges" while at work, and their exposure is carefully monitored. The health records of these occupationally exposed groups often show that they have lower rates of mortality from cancer and other causes than the general public and, in some cases, significantly lower rates than other workers who do similar work without being exposed to radiation.

mutations in fruit flies. Intensive study since then has shown that radiation can similarly induce mutations in plants and test animals. However evidence of genetic damage to humans from radiation, even as a result of the large doses received by atomic bomb survivors in Japan, **has not shown any such effects**.

In a plant or animal cell the material (DNA) which carries genetic information necessary to cell development, maintenance and division is the critical target for radiation. Much of the damage to DNA is repairable, but in a small proportion of cells the DNA is permanently altered. This may result in death of the cell or development of a cancer, or in the case of cells forming gonad tissue, alterations which continue as genetic changes in subsequent generations. Most such mutational changes are deleterious; very few can be expected to result in improvements.

The levels of radiation allowed for members of the public and for workers in the nuclear industry are such that any increase in genetic effects due to nuclear power will be imperceptible and almost certainly non-existent. Radiation exposure levels are set so as to prevent tissue damage and minimize the risk of cancer. Experimental evidence indicates that cancers are more likely than genetic damage. Some 75,000 children born of parents who survived high radiation doses at Hiroshima and Nagasaki in 1945 have been the subject of intensive examination. This study confirms that no increase in genetic abnormalities in human populations is likely as a result of even quite high doses of radiation. Similarly, no genetic effects are evident as a result of the Chernobyl accident.

Life on earth commenced and developed when the environment was certainly subject to several times as much radioactivity as it is now, so radiation is not a new phenomenon. If we ensure that there is no dramatic increase in people's general radiation exposure, it is most unlikely that genetic damage due to radiation will ever become significant.

7.5 REACTOR SAFETY

There have been sophisticated statistical studies on reactor safety. However, for most people actual performance is more convincing than probability statistics. **The situation to date is that in over 12,000 reactor-years of civil operation there has been only one accident to a commercial reactor which was not substantially contained within the design and structure of the reactor.** To this experience one could add another 12,000 reactor-years of naval operation, which in the west has had an excellent safety record.

Only the Chernobyl disaster in 1986 resulted in radiation doses to the public greater than those resulting from exposure to natural sources. Other incidents (and one "accident") have been completely confined to the plant. The tragedy made it clear why such reactors have never been licensed in other parts of the world. Apart from Chernobyl, no nuclear workers or members of the public have ever died as a result of exposure to radiation due to a commercial nuclear reactor incident. This is remarkable for the first five decades of a complex new technology which is being used in 30 countries, some reactors now operating having been built over forty years ago.

Most of the serious radiological injuries and deaths that occur each year (2-4 deaths and many more exposures above regulatory limits) are the result of large uncontrolled radiation sources, such as abandoned medical or industrial equipment. These have nothing to do with nuclear power generation.

Most accident scenarios involve primarily a loss of cooling. This may lead to the fuel in the reactor core overheating, melting and releasing fission products. Hence the provision of emergency core cooling systems on standby. In case these should fail, further protective barriers come into play: the reactor core is normally enclosed in structures designed to

120

prevent radioactive releases to the environment. Regulatory requirements today are that the effects of any core-melt accident must be confined to the plant itself, without the need to evacuate nearby residents. About one third of the capital cost of reactors is normally due to engineering designed to enhance the safety of people – both operators and neighbours, if and when things go wrong. Table 14 shows the international scale for reporting nuclear accidents or incidents.

The main safety concern has always been over the possibility of an uncontrolled release of radioactive material, leading to contamination and subsequent radiation exposure to people nearby. Earlier assumptions were that this would be likely in the event of a major loss-of-cooling accident which resulted in a core melt. Experience has proved otherwise in any circumstances relevant to Western reactor designs. In the light of better understanding of the physics and chemistry of material in a reactor core under extreme conditions, it became evident that even a severe core melt coupled with breach of containment could not in fact create a major radiological disaster from any Western reactor design. Studies of the post-accident situation at Three Mile Island in 1979, where there was no breach of containment, supported this. The total radioactivity release from this accident was small, and the maximum dose to individuals living near the power plant was well below internationally-accepted limits, even though the reactor was written off. Nevertheless, this accident had a pronounced psychological impact, was a severe blow to the US nuclear industry and had an adverse effect on the growth of nuclear capacity in the USA and beyond. More positively, it brought about profound changes in the way reactors are run, and in details of their engineering. In retrospect it was a very valuable stimulus to

improvements, and had much the same effect on reactor safety as the Comet airliner crashes of the 1950s did on the safety of pressurized jet aircraft – to everybody's benefit today.

The 1986 accident at **Chernobyl** in Ukraine was very serious due to the design of the reactor and its burning graphite, which dispersed radioactive contamination far and wide. It cost the lives of 47 staff and firefighters, 28 of them from acute radiation exposure. There have also been 1800 cases of thyroid cancer registered in children, most of which were curable, though about 10 have been fatal. No increase in leukaemia and other cancers had shown up in the first decade, but the World Health Organization (WHO) expects some increase in cancers over the next decade, and the death toll from delayed health effects may well climb beyond the ten or so thyroid cancer victims. About 130,000 people received significant radiation doses (i.e. above ICRP limits), and are being closely monitored by WHO. Radioactive pollution drifted across a wide area of Europe and Scandinavia, causing disruption to agricultural production and some exposure (small doses) to a large population[5].

The accident drew public attention to the lack of an adequate containment structure such as is standard on Western reactors. In addition, the RBMK design was such that coolant failure led to strong increase in power output from the fission process. Under abnormal conditions all reactor types may experience power increases, which are controlled by the reactor shutdown system and by the design physics. Light water reactors, in which the coolant serves as moderator, automatically reduce power when the coolant/moderator is lost, and can then be shut down using the control rods.

It has long been asserted that nuclear reactor accidents are the epitome of low-probability

---

[5] See: Chernobyl Ten Years On, OECD NEA 1996.

but high-consequence risks. However, the physics and chemistry of a reactor core, coupled with but not wholly depending on the engineering, mean that the consequences of an accident are likely in fact to be much less severe than those from other industrial and energy sources. Experience bears this out.

The Chernobyl accident was caused by a combination of design deficiencies and the violation of operating procedures resulting from an absence of a safety culture. With assistance from the West, significant safety improvements have been made to the 12 RBMK reactors in operation in Russia and Lithuania and the one potentially under construction in Russia. Russian reactor design has since been standardized on PWR types with containment structures.

Soon after the accident the destroyed Chernobyl 4 reactor was enclosed in a large concrete shell. The other three units on the site initially resumed operation, though they have since shut down, the last at the end of 2000.

An OECD expert report concluded that "the Chernobyl accident has not brought to light any new, previously unknown phenomena or safety issues that are not resolved or otherwise covered by current reactor safety programmes for commercial power reactors in OECD Member countries." A very positive outcome of the accident was creation of the World Association of Nuclear Operators (WANO), which enables the sharing of expertise and experience across the world.

See also: WNA information papers on Chernobyl Accident & Three Mile Island Accident.

There have been a number of accidents in experimental reactors and in one military plutonium-producing pile, including a number of core melts, but none of these has resulted in loss of life outside the actual plant or long-term environmental contamination. The following table (Table 15) of serious reactor accidents includes those in which fatalities have occurred, together with the most serious commercial plant accidents. The list probably corresponds to incidents rating 4 or higher on today's International Nuclear Event Scale (Table 14). It should be emphasized that a commercial-type reactor simply cannot under any circumstances explode like a nuclear bomb.

In an uncontained reactor accident, such as at Windscale (a military facility) in 1957 and at Chernobyl in 1986, the principal health hazard is from the spread of radioactive materials, notably volatile fission products, such as iodine-131 and caesium-137. These are biologically active, so that if consumed in food, they tend to stay in organs of the body. I-131 has a half-life of 8 days, so it is a hazard for around the first month, (and apparently gave rise to the thyroid cancers after the Chernobyl accident). Caesium-137 has a half-life of 30 years, and is therefore potentially a long-term contaminant of pastures and crops. In addition to these, there is caesium-134, which has a half-life of about two years. While measures can be taken to limit human uptake of I-131, (evacuation of area for several weeks, ingestion of iodine tablets), radioactive caesium can preclude food production from affected land for a long time. Other radioactive materials in a reactor core have been shown to be less of a problem because they are either not volatile (strontium, transuranic elements) or not biologically active (tellurium-132).

Despite the commercial nuclear power industry's impressive safety record and the thorough engineering of reactor structures and systems, which make a catastrophic radioactive release from any Western reactor extremely unlikely, there are those who simply don't want to run any risk of this. This fear must then be weighed against the benefits of nuclear power, in the same way that some people's fear of

Table 14: The international nuclear event scale.
For prompt communication of safety significance

Level, Descriptor	Off-site impact	On-site impact	Defence-in-depth degradation	Examples
7 Major Accident	*Major Release:* Widespread health and environmental effects			Chernobyl, USSR, 1986
6 Serious Accident	*Significant Release:* Full implementation of local emergency plans			Mayak at Ozersk, Russia, 1957 (reprocessing plant criticality)
5 Accident with Off-Site Risks	*Limited Release:* Partial implementation of local emergency plans	Severe core damage to reactor core or to radiological barriers		Windscale, UK, 1957 (military) Three Mile Island, USA, 1979 (fuel melting)
4 Accident Mainly in Installation either of:	*Minor Release:* Public exposure of the order of prescribed limits, or	Significant damage to reactor core or to radiological barriers, worker fatality		Saint-Laurent, France, 1980 (fuel rupture); Tokai Mura, Japan, 1999 (criticality in fuel plant for an experimental reactor)
3 Serious Incident any of:	*Very Small Release:* Public exposure at a fraction of prescribed limits, or	Major contamination. Acute health effects to a worker, or	Near accident. Loss of defence-in-depth provisions – no safety layers remaining	Vandellos, Spain, 1989 (turbine fire, no radioactive contamination) Davis-Besse, USA, 2002 (severe corrosion) Paks, Hungary 2003 (fuel damage)
2 Incident	nil	Significant spread of contamination. Overexposure of worker	Incidents with significant failures in safety provisions	
1 Anomaly	nil	nil	Anomaly beyond authorized operating regime	
0	nil	nil	No safety significance	
Below scale	nil	nil	No safety significance	

Source: International Atomic Energy Agency.

Table 15: Serious reactor accidents[6]

Serious accidents in military, research and commercial reactors. All except Browns Ferry and Vandellos involved damage to or malfunction of the reactor core. At Browns Ferry a fire damaged control cables and resulted in an 18-month shutdown for repairs; at Vandellos a turbine fire made the 17-year old plant uneconomic to repair.

Reactor	Date	Immediate deaths	Environmental effect	Follow-up action
NRX, Canada (experimental, 40 MWt)	1952	Nil	Nil	Repaired (new core). Closed 1992.
Windscale-1, UK (military plutonium-producing pile)	1957	Nil	Widespread contamination, farms affected (c. 1.5×10^{15} Bq released)	Entombed (filled with concrete). Being dismantled.
SL-1, USA (experimental, military, 3 MWt)	1961	Three operators	Very minor radioactive release	Decommissioned
Fermi-1 USA (experimental breeder, 66 MWe)	1966	Nil	Nil	Repaired and restarted, then closed in 1972
Lucens, Switzerland (experimental, 7.5 MWe)	1969	Nil	Very minor radioactive release	Decommissioned
Browns Ferry, USA (commercial, 2 x 1080 MWe)	1975	Nil	Nil	Repaired
Three Mile Island-2, USA (commercial, 880 MWe)	1979	Nil	Minor short-term radiation dose (within ICRP limits) to public; delayed release of 2×10^{14} Bq of Kr-85.	Clean-up programme complete. In "monitored storage" stage of decommissioning.
Saint Laurent-A2, France (commercial, 450 MWe)	1980	Nil	Minor radiation release (8×10^{10} Bq)	Repaired. Decommissioned 1992.
Chernobyl-4, Ukraine (commercial, 950 MWe)	1986	47 staff & fire-fighters, (32 immediate)	Major radiation release across E. Europe and Scandinavia (11×10^{18} Bq)	Entombed
Vandellos-1, Spain (commercial, 480 MWe)	1989	Nil	Nil	Decommissioned

[6] The well-publicized accident at Tokai Mura, Japan, in 1999, was at a fuel preparation plant for experimental reactors, and killed two workers from radiation exposure. Many other such criticality accidents have occurred, some fatal, and practically all in military facilities prior to 1980.

having aeroplanes crash on top of them must be balanced against the utility of air transport for the rest of the population. Ultimately, balancing risks and benefits is not simply a scientific exercise.

See also: WNA information paper on Cooperation in the Nuclear Power Industry.

Terrorism

Since the World Trade Centre attacks in New York in 2001, there has been concern about the consequences of a large aircraft being used to attack a nuclear facility with the purpose of releasing radioactive materials. Various studies have looked at similar attacks on nuclear power plants. They show that nuclear reactors would be more resistant to such attacks than virtually any other civil installations. A thorough study was undertaken by the Electric Power Research Institute in 2002, using specialist consultants and partly funded by the US Dept. of Energy. It concludes that US reactor structures "are robust and (would) protect the fuel from impacts of large commercial aircraft".

The analyses used a fully-fuelled Boeing 767-400 of over 200 tonnes as the basis, at 560 km/h – the maximum speed for precision flying near the ground. The wingspan of this aircraft is greater than the diameter of reactor containment buildings, and the 4.3 tonne engines are 15 metres apart. Hence analyses focused on single engine direct impact on the centre line and on the impact of the entire aircraft if the fuselage hit the centre line (in which case the engines would ricochet off the sides). In each case no part of the aircraft or its fuel would penetrate the containment.

Looking at spent fuel storage pools, similar analyses showed no breach. Dry storage and transport casks retained their integrity. "There would be no release of radionuclides to the environment".

Switzerland's Nuclear Safety Inspectorate studied a similar scenario and reported in 2003 that the danger of any radiation release from such a crash would be low for the older plants and extremely low for the newer ones.

Similarly, the massive structures mean that any terrorist attack even inside a plant (which are well defended) would not result in any significant radioactive releases.

The conservative design criteria which caused most power reactors to be shrouded by massive containment structures has provided peace of mind in a suicide terrorist context. Ironically and as noted earlier, with better understanding of what happens in a core melt accident inside, they are now seen to be not nearly as necessary in that accident mitigation role as was originally assumed.

8

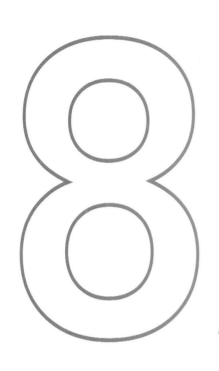

AVOIDING WEAPONS PROLIFERATION

Like many other technological innovations, nuclear technology has been from the outset ambiguous. Its initial development was military, during World War II. Two nuclear bombs made from uranium-235 and plutonium-239 were dropped on Japan's Hiroshima and Nagasaki respectively in August 1945, and these brought the long war to a sudden end. The immense and previously unimaginable power of the atom had been demonstrated. There was a large death toll, and survivors of the original blasts have suffered from a slightly increased incidence of cancer.

Then attention turned to civil applications. In the course of half a century, nuclear technology has enabled humankind to access a virtually unlimited source of energy at a time when constraints are arising on the use of fossil fuels. The question which frames this chapter is: **To what extent and in what ways does nuclear power generation contribute to or alleviate the risk from nuclear weapons?**

In the 1960s it was widely assumed that there would be 30 to 35 nuclear weapons states by the turn of the century. In fact there were eight – a tremendous testimony to the effectiveness of the Nuclear Non-Proliferation Treaty and its incentives both against weapons and for civil nuclear power.

8.1 INTERNATIONAL COOPERATION

Nuclear weapons are now in the possession of several nations[1], and during the "Cold War" (1950s to 1980s) there was a massive build-up of nuclear armaments, particularly by the USA and the Soviet Union. In the last 40 years there have been strenuous international efforts to dissuade other countries from joining the five declared nuclear weapons states. These efforts have been central to the role of one particular body, the International Atomic Energy Agency (IAEA), set up in 1957 by unanimous resolution of the United Nations.

One of the main functions of the IAEA is "to establish and administer safeguards designed to ensure that special fissionable and other materials... are not used in such a way as to further any military purpose." The IAEA endeavours to detect any diversion of nuclear material from peaceful nuclear activities to the manufacture of nuclear weapons or other nuclear explosive devices. Further, it attempts to deter any such diversion by its capacity for early detection. The IAEA also advises its members on the use of nuclear materials in non-military areas, such as agriculture, industry and medicine, and develops safety standards for nuclear power plants.

Photograph supplied by IAEA ImageBank

International Atomic Energy Agency headquarters in Vienna

---

[1] Weapons states are USA, UK, Russia, France, and China. Israel is described as a "threshold state", maintaining ambiguity about its nuclear status but generally considered to have nuclear weapons capability. South Africa declared and then voluntarily dismantled a clandestine nuclear weapons programme. India and Pakistan have demonstrated possession of nuclear weapons (notably through tests in 1998) but can only join the NPT if they, like South Africa, voluntarily renounce and dismantle their nuclear weapons. North Korea's precise status is unknown (as of early 2006) but has sought to develop nuclear weapons. Iran is ambiguous, in pursuing uranium enrichment on a significant scale without any evident commercial justification.

At the time the IAEA was being established, there was considerable concern that many countries would seek to develop or acquire nuclear weapons, just as they might upgrade their military forces with new equipment.

It was in this context that the cornerstone document governing the spread of nuclear weapons, the Treaty on the Non-proliferation of Nuclear Weapons (Non-Proliferation Treaty or NPT), was negotiated. **The NPT was essentially an agreement between the five nuclear weapons states and the other countries interested in nuclear technology. The deal was that assistance and cooperation would be traded for pledges, backed by international scrutiny, that no plant or material would be diverted to weapons' use.** Those who refused to be part of the deal would be excluded from international cooperation or trade involving nuclear technology. The NPT also represented a nuclear truce among non-weapons states, whereby they collectively resolved to turn away from the nuclear weapons option.

The first group of NPT signatories are non-nuclear weapon states. Each must agree not to manufacture or otherwise acquire nuclear weapons or other nuclear explosive devices. These states are obliged to conclude agreements with the IAEA for the application of safeguards on the full scope of their nuclear programme (see section 8.2).

The other NPT signatories are the five so-called nuclear weapons states. This group includes those who had manufactured and exploded a nuclear weapon before 1967, and consists of the USA, the Soviet Union (now Russia), the UK, France and China[2]. These countries are not required to accept IAEA safeguards, although the NPT does contain certain obligations concerning disarmament

Photograph supplied by IAEA ImageBank

Former IAEA Director General Hans Blix together with present incumbent Mohamed ElBaradei

which apply to them. All have, however, signed the NPT and accepted some safeguards on their peaceful nuclear activities.

The NPT entered into force in 1970 and was extended indefinitely in 1995. It is complemented by several regional treaties. Recently, other developments aimed at bolstering the non-proliferation regime have emerged. In September 1996 a Comprehensive Test Ban Treaty was opened for signature, aimed at the elimination of nuclear weapons testing. Negotiations are under way on a Fissile Material Cut-off Treaty, which would prohibit the further production of fissile nuclear weapons materials.

---

[2] France and the People's Republic of China did not ratify the NPT until 1992.

8.2 INTERNATIONAL NUCLEAR SAFEGUARDS

Over more than 35 years the IAEA's safeguards system under the NPT has been a conspicuous international success, at least within the scope of its operation. It has involved cooperation in developing nuclear energy for electricity generation, while ensuring that civil uranium, plutonium and associated plant did not allow weapons proliferation to occur as a result of this.

It is important to realize that international nuclear safeguards are focused on the control of fissile materials only. They have nothing to do with engineering or organizational safety aspects of reactors, waste disposal, or transport. These are covered by other international arrangements and conventions. It is also important to understand that nuclear safeguards are a prime means of reassurance whereby non-nuclear weapons states demonstrate to others that they are fulfilling their peaceful commitments. They prevent nuclear proliferation in the same way that auditing procedures build confidence in proper conduct and prevent embezzlement. Their specific objective is to verify whether declared (usually traded) nuclear material remains within the civil nuclear fuel cycle and is being used solely for peaceful purposes.

In other words, nuclear safeguards are intended to reveal whether a nation is adhering to its undertakings in relation to nuclear fuel materials. It is then up to the international community to bring pressure to bear on such a country if diversion of nuclear materials from its peaceful programme or other major irregularities are demonstrated. International nuclear safeguards are administered by the IAEA and were formally established under the NPT, which 187 states (including Taiwan) have signed. NPT safeguards require nations to:

- Declare to the IAEA their nuclear facilities
- Report to the IAEA what nuclear materials they hold and their location
- Accept visits by IAEA auditors and inspectors to verify independently their material reports and physically inspect the nuclear materials concerned, to confirm physical inventories of them

The IAEA also administers specific safeguards procedures for some countries[3] that have not joined the NPT. The IAEA safeguards are the principal nuclear control procedures in the world today, and cover almost 900 nuclear facilities and other locations containing nuclear material in 57 non-nuclear-weapons countries. However, other safeguards systems also exist, for example, amongst certain European nations (Euratom Safeguards) or between individual countries (bilateral agreements), such as Australia and customer countries for its uranium, or Japan and the USA.

These safeguards systems have been effective in preventing any diversion of materials actually covered by them. However, as nuclear power reactors, research reactors and fuel cycle components become more widespread, the safeguards task becomes more complex. At the same time more than simply accounting and audit is now expected of the safeguards, and concerns are focused on countries and activities not so far covered by them. Revision and upgrading of safeguards procedures is a continuing process.

For instance, Iraq showed up shortcomings in detection when it mounted an ambitious and clandestine indigenous weapons programme, which was unrelated to civil nuclear power. This provided the impetus for a thorough reconsideration of what safeguards are expected to achieve, and how they should be

---

[3] India, Pakistan, Israel, Cuba and Brazil. The first three have significant nuclear activities which are not subject to IAEA safeguards, although they accept them for some facilities. India has pledged to put all civil facilities under safeguards. Cuba and Brazil have all their nuclear activities under safeguards.

implemented beyond the normal trade in civil nuclear materials and related activities. The enhanced safeguards system resulting from this will provide a credible assurance that any undeclared nuclear activities would be detected in NPT countries. The focus of concern and political attention would then be squarely on countries defaulting on their international safeguards commitments (e.g. North Korea, Iraq and Iran) and on non-NPT countries, notably Israel, Pakistan and India.

Centrifuges that could be used to separate high-grade uranium from natural uranium, found in a warehouse near Tuwaitha, Iraq, after the first Gulf War

An example of improving safeguards is the agreement reached among nuclear exporting nations in the late 1970s, concerning restriction on sale of sensitive fuel cycle technologies (enrichment, fuel fabrication and reprocessing). Even Iraq found this difficult to get around. Nuclear reactor exports are also placed under tight control by this agreement, which stipulates the need for government assurances regarding peaceful use, together with acceptance by customer countries of full-scope safeguards inspections on all present and future nuclear activities. On the other hand, countries such as India, which developed its nuclear deterrent after the NPT came into effect (rather than just before, as China), and is thus denied any place within the NPT, is severely disadvantaged by the safeguards system in developing nuclear power for peaceful purposes. India's situation is now being addressed on the basis of its non-proliferation bona fides, and desire to be treated the same as China.

In May 1997 the IAEA started to develop and implement strengthened measures, now known as Integrated Safeguards, for use by the Agency when verifying states' compliance with their commitments not to produce nuclear weapons. This was in response to a widespread view that, having achieved so much in controlling trade in fissile materials, the IAEA could now look at any nuclear-related materials and technology as possible indicators of undeclared nuclear programmes and hence undeclared nuclear materials. It is hoped that most or all of the NPT's 187 signatories will eventually agree to these measures, which are detailed in an Additional Protocol to each country's agreement with IAEA, through which they would accept stronger and more intrusive verification on their territory. This has now become firmly established as the standard for NPT safeguards.

The new measures provide increased access for inspectors, both to information about current and planned nuclear programmes and to more locations on the ground. Access will not be restricted to declared nuclear sites, but will extend almost anywhere, including high-tech industrial facilities. Inspection activity may include remote surveillance, environmental sampling and monitoring systems at key locations. States accepting the protocol will need to remove restrictive requirements on inspectors so that they can visit anywhere at short notice. In practice, this is proving more of a deterrent to full take-up of the Additional Protocol than was initially envisaged.

Today many nations have the necessary trained scientists, experienced chemical technicians and the raw materials to attempt to carry out a moderate weapons production programme if they so desire, as Iraq demonstrated. Certainly the widespread use of nuclear power for electricity generation together with the large numbers of research reactors operating in over 50 countries has resulted in many people being trained and experienced in aspects of nuclear operations.

The most important factor underpinning the safeguards regime is international political pressure and how particular nations perceive their long-term security interests in relation to their immediate neighbours.

The solution to weapons proliferation is thus political more than technical, and it certainly goes beyond the question of uranium availability. International pressure not to acquire weapons is enough to deter most states from developing a weapons programme. The major risk of nuclear weapons proliferation will always lie with countries which have not joined the NPT and which have significant unsafeguarded nuclear activities, and those which have joined but disregard their treaty commitments. India, Pakistan and Israel are in the first category; North Korea, Iraq and Iran are in the second. While safeguards apply to some nuclear activities in non-NPT countries, others remain outside the safeguards' scrutiny.

8.3 FISSILE MATERIALS

Much of the concern about possible weapons proliferation arises from considering the fissile materials themselves. For instance, in relation to the plutonium contained in spent fuel discharged each year from the world's commercial nuclear power reactors, it is correctly but misleadingly asserted that "only a few kilograms of plutonium are required to make a bomb". Furthermore, no nation is without enough indigenous uranium to construct a few weapons (see section 3.3).

Table 16 gives some of the important characteristics of plutonium and its use. Plutonium is a substance of varying properties depending on its source. It consists of several different isotopes, including Pu-238, Pu-239, Pu-240, and Pu-241. All of these are "plutonium" but not all are fissile – only Pu-239 and Pu-241 can undergo fission in a normal reactor. Plutonium-239 by itself is an excellent nuclear fuel. It has also been used extensively for nuclear weapons because it has a relatively low spontaneous fission rate and a low critical mass. Consequently Pu-239, with only a few percent of the other isotopes present, is often called "weapons-grade" plutonium. This was used in the Nagasaki bomb in 1945, and in many of those in world weapons stockpiles since.

On the other hand, "reactor-grade" plutonium as routinely produced in all commercial nuclear power reactors, and which may be separated by reprocessing the used fuel from them, is not the same thing at all. It contains a large proportion – up to 40% – of the heavier plutonium isotopes, especially Pu-240, because it remained in the reactor for a relatively long time while much of the Pu-239 produced was burned up (see Figure 21). This composition is not a particular problem for re-use of the plutonium in mixed oxide (MOX) fuel for reactors (see sections 4.2 and 5.2), but it seriously affects the suitability of the material for nuclear weapons.

Table 16: Plutonium

Formation

U-238 + neutron $\Rightarrow$ U-239 $\Rightarrow$ Np-239 $\Rightarrow$ Pu-239

(beta decays of U-239 and Np-239: 23.5 min. and 2.35 days half-life respectively)

Pu-239 + neutron $\Rightarrow$ Pu-240

Pu-240 + neutron $\Rightarrow$ Pu-241

On average, one in four neutron absorptions by Pu-239 results in the formation of Pu-240 rather than in fission. Pu-241 and Pu-242 are formed by successive neutron capture in the reactor fuel. After fuel has been irradiated in the reactor for a couple of years, Pu-239 burns almost as fast as it forms, whereas Pu-240 accumulates steadily.

A very small amount of Pu-238 is formed from U-235 by neutron capture.

Amount

A 1000 MWe reactor produces about 250 kg of plutonium (especially Pu-239) each year. It remains locked up in highly radioactive spent fuel unless reprocessed (see Figure 16).

The amount of Pu-240 increases with the time that fuel elements remain in the reactor (see Figure 21). Pu-240 is not fissile in a thermal reactor, but can become fissile Pu-241 by further neutron capture. (Pu-240 is fissionable in a fast neutron reactor.)

Radioactivity:

Pu-239 emits alpha particles to decay to U-235 (see Appendix 2). Its half-life is 24,390 years, therefore it has a low level of radioactivity.

Pu-240 emits alpha particles as it decays to U-236 (another non-fissile isotope). Its half-life is 6600 years, therefore it has a higher level of radioactivity than Pu-239. It also emits neutrons from spontaneous fission disintegrations, as does Pu-238 (half-life 86 years).

Providing protection from this alpha radioactivity involves sealing the plutonium from physical contact, such as in a plastic bag.

Uses

The decay heat of Pu-238 (0.56 W/g) enables its use as an energy source in the thermoelectric generators of some cardiac pacemakers, space satellites, navigation beacons and so on. Plutonium power enabled the Voyager spacecraft to send back pictures of distant planets. Pu-240 has been used in similar applications.

The main peaceful use of Pu-239 is as nuclear reactor fuel.

Pu-241 (half-life 13 years) is the source, by beta decay, of americium-241, the vital ingredient in most household smoke detectors.

Type	Composition	Origin	Use
Reactor-grade, from high-burnup fuel	55%-60% Pu-239, >19% Pu-240, typically about 30% non-fissile	Comprises about 1% of spent fuel from normal operation of civil nuclear reactors used for electricity generation	As ingredient (c. 5%-7%) of MOX fuel for normal reactor (can also be used as fuel in fast neutron reactor)
Weapons-grade	Pu-239 with <7% Pu-240	From military "production" reactors specifically designed and operated for production of low burnup Pu	Nuclear weapons (can be recycled as fuel in fast neutron reactor or as ingredient of MOX)

Due to the spontaneous fission of Pu-240, only a very low level of it is tolerable in material for making weapons. Design and construction of nuclear explosives based on normal reactor-grade plutonium would be difficult, dangerous and unreliable, and has not so far been done[4]. However, safeguards arrangements assume that both kinds of plutonium could conceivably be used for weapons, particularly weapons designed for terror rather than military use. This is the basis of objection from some quarters to reprocessing and separation of any plutonium from used fuel.

It is worth noting that a nuclear reactor which uses mixed oxide input for one third of its fuel is not a net producer of plutonium, and that which emerges in the fuel is even less suitable for weapons use than what is in the fresh MOX fuel.

Commercial plutonium is therefore a very much less attractive material for weapons than plutonium produced in special "production reactors"[5] designed for producing Pu-239, which are, moreover, capable of frequent fuel changing. However, the development of laser enrichment technology may mean that it becomes feasible to enrich commercial plutonium to weapons grade. Hence safeguards arrangements are set up accordingly to take seriously the proliferation possibilities even of reactor-grade plutonium. (Conventional enrichment cannot readily be used to separate Pu-239 from Pu-240 because the atomic mass is so similar.)

The plutonium-based fast breeder fuel cycle is seen as having features which might give rise to weapons proliferation problems. On the other hand, conventional thermal reactors normally have a higher net yield of plutonium from the fuel cycle (see Figures 13 and 14). This suggests that in the foreseeable future the fast neutron reactor should be utilized more as a plutonium "incinerator" (see section 4.4), which is likely to mean less plutonium in storage or in spent fuel elements than otherwise.

Figure 21: Plutonium in the reactor core

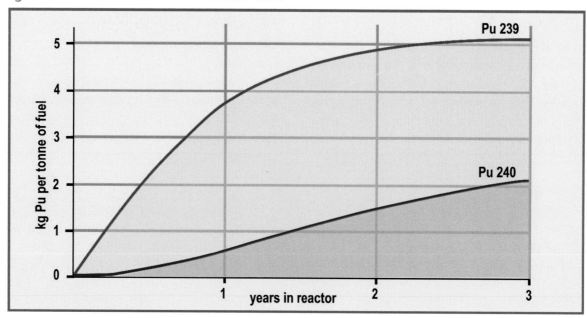

[4] In 1962 a nuclear device using low-burnup plutonium from a UK power reactor was detonated in the USA. The isotopic composition of this plutonium has not been disclosed, but it was evidently about 90% Pu-239.
[5] Or heavy water-moderated research reactors, as in India.

There are two other fissile materials that could be used for weapons, and both are isotopes of uranium. The most common, and the material used to make the 1945 Hiroshima bomb, is uranium-235. This material is produced by enriching natural uranium in an enrichment plant, not to 3% or 4% as required for light water reactor fuel, but to 93% U-235 or higher.

The other isotope of uranium suitable for use in explosives is U-233. This material is made from thorium-232 fuels in (special) reactors in much the same way as plutonium is made from U-238 in uranium-fuelled reactors (see section 4.2). However, the use of thorium-fuelled reactors (see section 3.6) has not moved beyond the experimental stage, and U-233 is not seen as a significant proliferation problem.

Whilst the above materials can be used for explosives manufacture, they are not readily available in any practical sense, and international efforts are designed to make them even less accessible.

In recent years, the international community was challenged by an illicit nuclear weapons programme in North Korea, based on plutonium production in a research reactor and detected by IAEA safeguards inspections. The United Nations imposed a nuclear "freeze" on the country's reactors and facilities under a 1994 Agreed Framework, which led to the country bowing to international pressure so that the IAEA could reassure the UN that all nuclear materials were safeguarded and that North Korea was moving towards full compliance with its IAEA safeguards agreement. The trade-off for North Korea was that an international consortium led by the USA, South Korea and Japan started building two large modern nuclear reactors for the country to provide electricity untainted by military possibilities. However, in 2002 North Korea admitted to a clandestine uranium enrichment programme, which put the country doubly in default of its international treaty obligations. It then left the NPT, placing itself outside the IAEA safeguards regime.

Even greater concern was generated by suspicions that Iraq had developed or was developing nuclear weapons; these fears were heightened during the 1991 Persian Gulf War. After the cease-fire in 1991, the United Nations was able to confirm that Iraq, though a signatory to the NPT, had been pursuing a clandestine weapons programme quite separately from materials and facilities covered by IAEA inspections. The major part of its illicit endeavour was based on indigenous uranium and its enrichment. As noted in section 8.2, this situation led to the enhancement of the safeguards regime, through the IAEA's Integrated Safeguards Programme.

Questions continue as to the nuclear military intentions of India, Pakistan and Israel. None of these countries are bound by the NPT. India and Pakistan have demonstrated their capability of producing nuclear weapons, and Israel is suspected of having developed a nuclear weapons programme. Pakistan was found to have supplied material and technology to Iran, Libya and North Korea.

Iran attracted world attention in 2002 when previously undeclared nuclear facilities became the subject of IAEA inquiry. On investigation, the IAEA found inconsistencies in Iran's declarations to the Agency and has raised questions as to whether Iran was in violation of its safeguards agreement, as a signatory of the NPT. An IAEA report released to its member states in November 2003 showed that Iran had, in a series of contraventions of its safeguards agreement over 22 years, systematically concealed its development of key techniques – notably uranium enrichment – which are capable of use for nuclear weapons. The situation remained unresolved as of spring 2006.

8.4 RECYCLING MILITARY URANIUM AND PLUTONIUM FOR ELECTRICITY

International efforts aimed at nuclear disarmament have, ironically, led to some serious safety and security problems. Dismantling of nuclear warheads under USA-Russia disarmament agreements (START I and START II) has resulted in an accumulation of weapons-grade material (plutonium and high-enriched uranium). Concerns have arisen, particularly following the break-up of the Soviet Union, about the possibility that these fissile materials could be subject to theft, smuggling, illicit trafficking, and could make their way into the hands of rogue states or terrorists. Inadequate control of nuclear materials inside Russia, the sheer size of Russian nuclear programmes, and substandard security at nuclear installations are a few of the factors that have raised concerns about nuclear materials falling into the wrong hands. The joint efforts of many nations through the 1990s improved the physical security and accountability of such materials considerably.

The challenge of isolating and disposing of weapons-grade fissile material, particularly plutonium, that is no longer required for military purposes has therefore become a priority for the international community. The IAEA has been examining policy options concerning the management and use of stocks of military plutonium. The most pressing concern is its protection from theft and diversion, while determining the most appropriate means of disposition[6].

The prospect of using weapons-grade plutonium (more than 93% Pu-239) in mixed

Photograph supplied by IAEA ImageBank

oxide (MOX) fuel for civil reactors is receiving increased attention. It would be quite feasible to make MOX using a mixture of military and reactor-grade plutonium with depleted uranium, and some has been made with military plutonium only. MOX would be the only means of disposal which permanently removes military plutonium from circulation and effectively destroys it. Efforts are currently under way to "recycle" plutonium in this manner, and the so-called G-8 countries are exploring this option further.

After three decades of concern regarding the possibility of uranium intended for commercial nuclear power finding its way into weapons, we are now seeing military uranium being directed into the civil nuclear fuel cycle for use in commercial nuclear power generation. The first such material from Soviet military warheads arrived in the USA in 1995, and it now provides 10% of all US electricity. A start has also been made on recycling US weapons-grade uranium for electricity. Military high-enriched uranium is diluted about 25:1 with either depleted uranium left over from enrichment plants or similar material (see also section 3.5).

[6] The production of reactor-grade plutonium in spent fuel from civil reactors, at almost 100 tonnes per year, far exceeds that of weapons-grade plutonium. However, most is not separated from spent fuel, and if it has any weapons potential at all, it is minor compared with the weapons-grade material.

8.5 AUSTRALIAN AND CANADIAN NUCLEAR SAFEGUARDS POLICIES

Canada and Australia produce over half of the world's mined uranium and therefore provide a case study in how non-proliferation is approached. Both countries are strong proponents of a robust international non-proliferation regime to enhance national and international security. Both are rigorous in seeking assurances that nuclear exports will only be used for legitimate and peaceful nuclear energy purposes.

Beyond that, Australia's main interest in international nuclear safeguards is in relation to the use of its uranium in overseas nuclear power programmes. Canada's interest is broader, covering the whole domestic fuel cycle, plus the export of both uranium and reactor technology. In both countries, exports of uranium are controlled by the federal governments.

Following World War II, Canada pledged that it would not develop nuclear weapons, even though it had, at the time, the capability to do so. Both Canada and Australia participated in the drafting of the Statute of the IAEA, have been continuously represented on the IAEA's Board of Governors, and remain active in many of the various technical committees and advisory groups of the IAEA.

In Australia the Ranger Uranium Environmental Inquiry commissioners pointed out quite clearly in their first report (1976) the importance of adequate safeguards being applied to Australia's uranium. The Australian Government then decided on the basic principles of an Australian safeguards policy, and these were announced during 1977. Australia was involved in the International Nuclear Fuel Cycle Evaluation Programme in the 1970s and continues to use its status as a uranium supplier to press for high safeguards standards to be applied. In so doing,

Table 17: Australian and Canadian nuclear safeguards policies

1. **Selected countries**

    Non-weapons states must be party to NPT and must accept full-scope IAEA safeguards applying to all their nuclear-related activities. Australia requires them to have ratified the Additional Protocol to their safeguards agreement with the IAEA.

    Weapons states to give assurance of peaceful use; IAEA safeguards to cover the material.

2. **Bilateral agreements are required**

    IAEA to monitor compliance with IAEA safeguards requirements
    Fallback safeguards (if NPT ceases to apply or IAEA cannot perform its safeguards functions)
    Prior consent to transfer material or technology to another country
    Prior consent to enrich above 20% U-235
    Prior consent to reprocess
    Control over storage of any separated plutonium
    Adequate physical security

3. **Materials exported or re-exported to be in a form attracting full IAEA safeguards.**

4. **Commercial contracts to be subject to conditions of bilateral agreements.**

5. **Australia and Canada will participate in international efforts to strengthen safeguards.**

6. **Australia and Canada recognize the need for constant review of standards and procedures.**

McArthur River underground mine – the world's largest high-grade uranium deposit

Australia is allied with Canada, the Western world's largest uranium producer.

Table 17 sets out in summary the main elements of both countries' policies.

The Australian and Canadian policies as outlined are based on the requirements of the Nuclear Non-Proliferation Treaty (NPT) and the IAEA safeguards invoked under it. Superimposed on these are conditions which are required by bilateral agreement with customer countries[7] and implemented by the Australian Safeguards and Non-proliferation Office (ASNO) or the Canadian Nuclear Safety Commission (CNSC) respectively.

Both countries' legally-binding bilateral safeguards measures are directed towards preventing any unauthorized or clandestine use of exported uranium or any materials derived from it: "Australian-obligated nuclear materials" or the Canadian equivalent. The Canadian agreements cover nuclear material, heavy water, nuclear equipment and technology. The bilateral safeguards are designed to deter possible diversion of fissile material or misuse of equipment and technology more effectively than standard IAEA safeguards on their own.

The Canadian federal nuclear regulatory agency is the Canadian Nuclear Safety Commission. The CNSC is responsible for regulating domestic nuclear facilities and is charged with administering the agreement between Canada and the IAEA for the application of safeguards in Canada. The commission assists the IAEA by allowing access to Canadian nuclear facilities and arranging for the installation of safeguards equipment at the sites. It also reports regularly to the IAEA on nuclear materials held in Canada. The CNSC also manages a programme for research and development in support of IAEA safeguards, the Canadian Safeguards Support Programme.

In Australia the Australian Safeguards and Non-proliferation Office performs a similar role, apart from the regulation. It administers the safeguards agreement with the IAEA, arranges IAEA access to Australian facilities, and reports to the IAEA on nuclear materials in Australia. ASNO also manages the Australian Safeguards Assistance Programme.

See also: WNA information paper on Safeguards to prevent proliferation.

---

[7] Australia has 18 bilateral safeguards agreements covering 36 countries (the Euratom agreement covering 25); Canada has 20 agreements in force, including with Euratom.

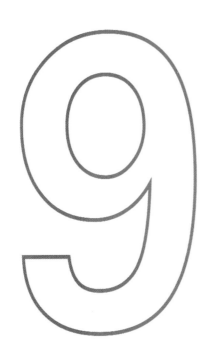

9

HISTORY OF NUCLEAR ENERGY

Uranium was discovered in 1789 by Martin Klaproth, a German chemist, and named after the planet Uranus.

9.1 EXPLORING THE NATURE OF THE ATOM

Ionizing radiation was discovered by Wilhelm Röntgen in 1895, by passing an electric current through an evacuated glass tube and producing continuous X-rays. Then in 1896 Henri Becquerel found that pitchblende (an ore containing radium and uranium) caused a photographic plate to darken. He went on to demonstrate that this was due to beta radiation (electrons) and alpha particles (helium nuclei) being emitted. Paul Villard found a third type of radiation from pitchblende: gamma rays, which were much the same as X-rays. Then in 1896 Pierre and Marie Curie gave the name "radioactivity" to this phenomenon, and in 1898 isolated polonium and radium from the pitchblende. Radium was later used in medical treatment. In 1898 Samuel Prescott showed that radiation destroyed bacteria in food.

The science of atomic radiation, atomic change and nuclear fission was developed from 1895 to 1945, much of it in the last six of those years.

In 1902 Ernest Rutherford showed that radioactivity as a spontaneous event involving the emission of an alpha or beta particle from the nucleus created a different element. In 1919 he fired alpha particles from a radium source into nitrogen and found that nuclear rearrangement was occurring, with formation of oxygen. Niels Bohr advanced our understanding of the way electrons were arranged around the atom's nucleus through to the 1940s.

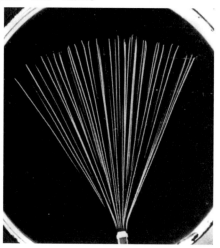

Early photo of alpha particles

By 1911 Frederick Soddy had discovered that naturally-radioactive elements had a number of different isotopes (radionuclides), with the same chemistry. Also in 1911, George de Hevesy showed that such radionuclides were invaluable as tracers, because minute amounts could readily be detected with simple instruments.

In 1932 James Chadwick discovered the neutron. In the same year, John Cockcroft and Ernest Walton produced nuclear transformations by bombarding atoms with accelerated protons. Then in 1934 Irene Curie and Frederic Joliot found that some such transformations created artificial radionuclides. The next year Enrico Fermi found that a much greater variety of artificial radionuclides could be formed when neutrons were used instead of protons.

Fermi continued his experiments, mostly producing heavier elements from his targets, but also, with uranium, some much lighter ones. In 1939 Otto Hahn and Fritz Strassman in Berlin showed that the new lighter elements were barium and others which were about half the mass of uranium, thereby demonstrating that atomic fission had occurred. Lise Meitner and her nephew Otto Frisch, working under Niels Bohr, then explained this by suggesting that the neutron was captured by the nucleus, causing severe vibration leading to the nucleus splitting into two not quite equal parts. They calculated the energy release from this fission as about 200 million electron volts. Frisch then confirmed this figure experimentally.

This was the first experimental confirmation of Albert Einstein's paper putting forward the equivalence between mass and energy, which had been published in 1905.

Photograph supplied by Niels Bohr Archive

9.2 HARNESSING NUCLEAR FISSION

The developments in 1939 sparked activity in many laboratories. Hahn and Strassman showed that fission not only released a lot of energy but that it also released additional neutrons which could cause fission in other uranium nuclei and possibly a self-sustaining chain reaction leading to an enormous release of energy. This suggestion was soon confirmed experimentally by Joliot and his co-workers in Paris, and by Leo Szilard working with Fermi in New York.

Bohr soon proposed that fission was much more likely to occur in U-235 than in U-238 and that fission would occur more effectively with slow-moving neutrons than with fast neutrons. The latter point was confirmed by Szilard and Fermi, who proposed using a "moderator" to slow down the emitted neutrons. Bohr and John Wheeler extended these ideas into what became the classical analysis of the fission process, and their paper was published only two days before war broke out in 1939.

Another important factor was that U-235 was known to comprise only 0.7% of natural uranium, with the other 99.3% being U-238; the two isotopes had the same chemical properties. Hence the separation of the two to obtain pure U-235 would be difficult and would require the use of their very slightly different physical properties. This increase in the proportion of the U-235 isotope became known as "enrichment".

The remaining piece of the fission/atomic bomb concept was provided in 1939 by Francis Perrin, who introduced the concept of the critical mass of uranium required to produce a self-sustaining release of energy. His theories were extended by Rudolph Peierls at Birmingham University, and the resulting calculations were of considerable importance in the development of the atomic bomb. Perrin's group in Paris continued their studies and demonstrated that a chain reaction could be sustained in a uranium-water mixture (the water being used to slow down the neutrons) provided external neutrons were injected into the system. They also demonstrated the idea of introducing neutron-absorbing material to limit the multiplication of neutrons and thus control the nuclear reaction (which is the basis for the operation of a nuclear power station).

Over 1939 to 1945, most development was focused on the atomic bomb.

Courtesy of Inst. Int. de Physique Solvay

7th Solvay conference in Brussels in 1933. Persons: Kramers, Hendrik Anton; Mott, Neville Francis; Gamow, George; Blackett, Patrick Maynard Stuart; Cosyns, M.; Piccard, Aug.; Stahel, E.; Dirac, Paul Adrian Maurice; Errera, J.; Ellis, Charles Drummond; Lawrence, Ernest Orlando; Henriot, E.; Joliot-Curie, Frederic; Heisenberg, Werner Karl; Walton, E.T.S.; Debye, Peter; Cabrera, B.; Bothe, Walther William; Bauer, H.E.G.; Verschaffelt, J.E.; Cockcroft, John Douglas; Rosenfeld, Lèon; Perrin, F.; Fermi, Enrico; Rosenblum, M. Salomon; Pauli, Wolfgang; Herzen, E.; Peierls, Rudolf Ernst; Schrödinger, Erwin; Joliot-Curie, Irene; Bohr, N.; Joffe, Abram Feodorovich; Curie, Marie; Richardson, Owen Williams; Rutherford, Ernest; Broglie, Maurice de; Meitner, Lise; Chadwick, James; Langevin, Paul; Donder, Th. de; Broglie, Louis Victord.

9.3 NUCLEAR PHYSICS IN RUSSIA

Russian nuclear physics predates the Bolshevik Revolution by more than a decade. Work on radioactive minerals found in central Asia began in 1900, and the St Petersburg Academy of Sciences began a large-scale investigation in 1909. The 1917 Revolution gave a boost to scientific research, and over ten physics institutes were established in major Russian towns, particularly St Petersburg, in the years which followed. In the 1920s and early 1930s many prominent Russian physicists worked abroad, encouraged by the new regime initially as the best way to raise the level of expertise quickly. These included Kirill Sinelnikov, Pyotr Kapitsa and Vladimir Vernadsky.

By the early 1930s there were several research centres specializing in nuclear physics. Sinelnikov returned from Cambridge in 1931 to organize a department at the Ukrainian Physico-Technical Institute (FTI) in Kharkov, which had been set up in 1928. Academician Abram Ioffe formed another group at Leningrad FTI (including the young Igor Kurchatov), which in 1933 became the Department of Nuclear Physics under Kurchatov with four separate laboratories.

By the end of the decade, there were cyclotrons installed at the Radium Institute in Leningrad and the Leningrad FTI (the biggest in Europe). But by this time many scientists were beginning to fall victim to Stalin's purges – half the staff of Kharkov FTI, for instance, was arrested in 1939. Nevertheless, 1940 saw great advances being made in the understanding of nuclear fission including the possibility of a chain reaction. At the urging of Kurchatov and his colleagues, the Academy of Sciences set up a "Committee for the Problem of Uranium" in June 1940, chaired by Vitaly Khlopin, and a fund was established to investigate the central Asian uranium deposits. With Germany's invasion of Russia in 1941, much of the research switched to potential military applications.

Three leading Soviet nuclear physicists (from left to right) Abram Ioffe, Abram Alixanov, Igor Kurchatov

Photograph supplied by RIA Novosti

9.4 CONCEIVING THE ATOMIC BOMB

In Britain the refugee physicists Peierls and Otto Frisch (who stayed in England with Peierls after the outbreak of war) gave a major impetus to the concept of an atomic bomb in a three-page document known as the Frisch-Peierls Memorandum. In this they predicted that an amount of about 5 kg of pure U-235 could make a very powerful atomic bomb equivalent to several thousand tonnes of dynamite. They also suggested how such a bomb could be detonated, how the U-235 could be produced, and what the radiation effects might be in addition to the explosive effects. They proposed thermal diffusion as a suitable method for separating the U-235 from the natural uranium. This memorandum stimulated a considerable response in Britain at a time when there was little interest in the USA.

A group of eminent scientists known as the MAUD Committee was set up in Britain and supervised research at the Universities of Birmingham, Bristol, Cambridge, Liverpool and Oxford. The chemical problems of producing gaseous compounds of uranium and pure uranium metal were studied at Birmingham University and by Imperial Chemical Industries (ICI). Dr Philip Baxter[1] at ICI made the first small batch of gaseous uranium hexafluoride for Professor James Chadwick in 1940. ICI received a formal contract later in 1940 to make 3 kg of this vital material for the future work. Most of the other research was funded by the universities themselves.

Rudolph Peierls –
co-author of the Frisch-Peierls Memorandum

Two important developments came from the work at Cambridge. The first was experimental proof that a chain reaction could be sustained with slow neutrons in a mixture of uranium oxide and heavy water (i.e. the output of neutrons was greater than the input). The second owed itself to work by Egon Bretscher and Norman Feather and was based on earlier research by Hans Halban and Lew Kowarski soon after they arrived in Britain from Paris. When U-235 and U-238 absorb slow neutrons, the probability of fission in U-235 is much greater than in U-238. The U-238 is more likely to form a new isotope U-239, and this isotope rapidly emits an electron to become a new element with a mass of 239 and an Atomic Number of 93. This element also emits an electron and becomes a new element of mass 239 and Atomic Number 94, which has a much greater half-life. Bretscher and Feather argued on theoretical grounds that element 94 would be readily fissionable by slow and fast neutrons, and had the added advantage that it was chemically different to uranium and therefore could easily be separated from it.

This new development was also confirmed in independent work by Edwin McMillan and Philip Abelson in the USA in 1940. Nicholas Kemmer of the Cambridge team proposed the names neptunium and plutonium for the new elements 93 and 94 by analogy with the outer planets Neptune and Pluto beyond Uranus (uranium, element 92). The Americans fortuitously suggested the same names, and the identification of plutonium in 1941 is generally credited to Glenn Seaborg.

---

[1] Dr Baxter later was sent to the Oak Ridge Laboratory in the USA to assist in the operation of the large enrichment plant, secretly constructed to make the material for the first atomic bombs. He later became a key figure in the Australian Atomic Energy Commission.

9.5 DEVELOPING THE CONCEPTS

By the end of 1940 remarkable progress had been made by the several groups of scientists coordinated by the MAUD Committee and for the expenditure of a relatively small amount of money. All of this work was kept secret, whereas in the USA several publications continued to appear in 1940, and there was also little sense of urgency.

By March 1941 one of the most uncertain pieces of information was confirmed – the fission cross-section of U-235. Peierls and Frisch had initially predicted in 1940 that almost every collision of a neutron with a U-235 atom would result in fission, and that both slow and fast neutrons would be equally effective. It was later discerned that slow neutrons were very much more effective, which was of enormous significance for nuclear reactors but fairly academic in the bomb context. Peierls then stated that there was now no doubt that the whole scheme for a bomb was feasible provided highly enriched U-235 could be obtained. The predicted critical size for a sphere of U-235 metal was about 8 kg, which might be reduced by use of an appropriate material for reflecting neutrons. However, direct measurements on U-235 were still necessary and the British pushed for urgent production of a few micrograms.

The final outcome of the MAUD Committee was two summary reports in July 1941. One was on "Use of Uranium for a Bomb", and the other was on "Use of Uranium as a Source of Power". The first report concluded that a bomb was feasible, and that one containing some 12 kg of active material would be equivalent to 1800 tons of TNT; it would moreover release large quantities of radioactive

James Chadwick: Discoverer of the neutron and member of the MAUD Committee

Photograph supplied by Cavendish Laboratory

substances, which would make places near the explosion site dangerous to humans for a long period. It estimated that a plant to produce 1 kg of U-235 per day would cost £5 million and would require a large skilled labour force that was also needed for other parts of the war effort. Suggesting that the Germans could also be working on the bomb, it recommended that the work should be continued with high priority in cooperation with the Americans, even though they seemed to be concentrating on the future use of uranium for power and naval propulsion.

The second MAUD Report concluded that the controlled fission of uranium could be used to provide energy in the form of heat for use in machines in industrial applications, as well as providing large quantities of radioisotopes, which could be used as substitutes for radium. It referred to the use of heavy water and possibly graphite as moderators for the fast neutrons, and that even ordinary water could be used if the uranium was enriched in the U-235 isotope. It concluded that the "uranium boiler" had considerable promise for future peaceful uses but that it was not worth considering during the present war. The Committee recommended that Halban and Kowarski should move to the USA, where there were plans to make heavy water on a large scale. The possibility that the new element plutonium might be more suitable than U-235 was mentioned, and that therefore the work in this area by Bretscher and Feather should be continued in Britain.

The two reports led to a complete reorganization of work on the bomb and the

Photograph supplied by Enrico Fermi Institute

Enrico Fermi: Directed the team which produced the first controlled nuclear chain reaction in 1942.

9.6 THE MANHATTAN PROJECT

The Americans increased their effort rapidly and soon outstripped the British. Research continued in each country with some exchange of information. Several of the key British scientists visited the USA early in 1942 and were given full access to all of the information available. The Americans were pursuing three enrichment processes in parallel: Ernest Lawrence was studying electromagnetic separation at Berkeley (University of California), E. V. Murphree of Standard Oil was studying the centrifuge method developed by Jessie Beams, and Harold Urey was coordinating the gaseous diffusion work at Columbia University. Responsibility for building a reactor to produce fissile plutonium was given to Arthur Compton at the University of Chicago. The British were only examining gaseous diffusion.

In June 1942 the US Army took over process development, engineering design, procurement of materials and site selection for pilot plants for four methods of making fissionable material (because none of the four had been shown to be clearly superior at that point) as well as the production of heavy water. With this change, information flow to Britain dried up. This was a major setback to the British and the Canadians who had been collaborating on heavy water production and on several aspects of the research programme. Thereafter, Churchill sought information on the cost of building a diffusion plant, a heavy water plant and an atomic reactor in Britain.

"boiler". It was claimed that the work of the committee had put the British in the lead and that "in its fifteen months' existence it had proved itself one of the most effective scientific committees that ever existed". The basic decision that the bomb project would be pursued urgently was taken by the Prime Minister, Winston Churchill, with the agreement of the Chiefs of Staff.

The reports also led to high level reviews in the USA, particularly by a Committee of the National Academy of Sciences, initially concentrating on the nuclear power aspect. Little emphasis was given to the bomb concept until 7 December 1941, when the Japanese attacked Pearl Harbour and the Americans entered the war directly. The huge resources of the USA were then applied without reservation to developing atomic bombs.

After many months of negotiations an agreement was finally signed by Churchill and President Roosevelt in Quebec in August 1943, according to which the British handed over all of their reports to the Americans and in return received copies of General Groves' progress reports to the President. The latter showed that the entire US programme would cost over $1000 million – all for the bomb, as no work was being done on other applications of nuclear energy.

Construction of production plants for electromagnetic separation (in calutrons) and gaseous diffusion was well under way. An experimental graphite pile constructed by Fermi had operated at the University of Chicago in December 1942 – the first controlled nuclear chain reaction.

A full-scale production reactor for plutonium was being constructed at Argonne, with further ones at Oak Ridge and then Hanford, plus a reprocessing plant to extract the plutonium. Four plants for heavy water production were being built, one in Canada and three in the USA. A team under Robert Oppenheimer at Los Alamos in New Mexico was working on the design and construction of both U-235 and Pu-239 bombs. The outcome of the huge effort, with assistance from the British teams, was that sufficient Pu-239 and highly enriched U-235 (from calutrons and diffusion at Oak Ridge) was produced by mid 1945. The uranium mostly originated from the Belgian Congo.

The first atomic device tested successfully at Alamagordo, New Mexico, on 16 July 1945. It used plutonium made in a nuclear pile. The teams did not consider that it was necessary to test a simpler U-235 device. The first atomic bomb, which contained U-235, was dropped on Hiroshima on 6 August 1945. The second bomb, containing Pu-239, was dropped on Nagasaki on 9 August. That same day, the USSR declared war on Japan. On 10 August 1945 the Japanese Government surrendered.

9.7 THE SOVIET BOMB

Initially Stalin was not enthusiastic about diverting resources to develop an atomic bomb, until intelligence reports suggested that such research was under way in Germany, Britain and the USA. Consultations with Academicians Abram Ioffe, Pyotr Kapitsa, Vitaly Khlopin and Vladimir Vernadsky convinced him that a bomb could be developed relatively quickly, and he initiated a modest research programme in 1942. Igor Kurchatov, then relatively young and unknown, was chosen to head it, and in 1943 he became Director of Laboratory No. 2, recently established on the outskirts of Moscow. This was later renamed LIPAN, then became the Kurchatov Institute of Atomic Energy. Overall responsibility for the bomb programme rested with Security Chief Lavrenti Beria and its administration was undertaken by the First Main Directorate (later called the Ministry of Medium Machine Building).

Research had three main aims: to achieve a controlled chain reaction, to investigate methods of isotope separation and to look at designs for both enriched uranium and plutonium bombs. Attempts were made to initiate a chain reaction using two different types of atomic pile: one with graphite as a moderator and the other with heavy water. Three possible methods of isotope separation were studied: counter-current thermal diffusion, gaseous diffusion and electromagnetic separation.

After the defeat of Nazi Germany in May 1945, German scientists were "recruited" to the bomb programme to work in particular on isotope separation to produce enriched uranium. This included research into gas centrifuge technology in addition to the three other enrichment technologies.

The test of the first US atomic bomb in July

1945 had little impact on the Soviet effort, but by this time, Kurchatov was making good progress towards both a uranium and a plutonium bomb. He had begun to design an industrial scale reactor for the production of plutonium, while those scientists working on uranium isotope separation were making advances with the gaseous diffusion method.

It was the bombing of Hiroshima and Nagasaki the following month

Replica of the first Soviet atomic bomb in the Russian Federal Nuclear Centre Museum

Photograph supplied by RIA Novosti

which gave the programme a high profile, and construction began in November 1945 of a new city in the Urals which would house the first plutonium production reactors – Chelyabinsk-40 (later known as Chelyabinsk-65 or the Mayak Production Association). This was the first of ten secret nuclear cities to be built in the Soviet Union. The first of five reactors at Chelyabinsk-65 came on line in 1948. This town also housed a processing plant for extracting plutonium from irradiated uranium.

As for uranium enrichment technology, it was decided in late 1945 to begin construction of the first gaseous diffusion plant at Verkh-Neyvinsk (later the closed city of Sverdlovsk-44), some 50 kilometres from Yekaterinburg (formerly Sverdlovsk) in the Urals. Special design bureaux were set up at the Leningrad Kirov Metallurgical and Machine-Building Plant and at the Gorky (Nizhny Novgorod) Machine-Building Plant. Support was provided by a group of German scientists working at the Sukhumi Physical Technical Institute.

In April 1946 design work on the bomb was shifted to Design Bureau-11 – a new centre at Sarova, some 400 kilometres from Moscow (subsequently the closed city of Arzamas-16). More specialists were brought in to the programme, including metallurgist Yefim Slavsky, who was given the immediate task of producing the very pure graphite Kurchatov needed for his plutonium production pile constructed at Laboratory No. 2, known as F-1. The pile was operated for the first time in December 1946. Support was also given by Laboratory No. 3 in Moscow – now the Institute of Theoretical and Experimental Physics – which had been working on nuclear reactors.

Work at Arzamas-16 was influenced by foreign intelligence gathering, and the first device was based closely on the Nagasaki bomb (a plutonium device). In August 1947 a test site was established near Semipalatinsk in Kazakhstan and was ready for the detonation two years later of the first bomb, RSD-1. Even before this was tested in August 1949, another group of scientists led by Igor Tamm and including Andrei Sakharov had begun work on a hydrogen bomb.

9.8 REVIVAL OF THE "NUCLEAR BOILER"

By the end of World War II, the project predicted and described in detail only five and a half years before in the Frisch-Peierls Memorandum had been brought to partial fruition, and attention could now turn to the peaceful and directly beneficial application of nuclear energy. Post-war, weapons development continued on both sides of the "iron curtain", but a new focus was on harnessing atomic power, now dramatically (if tragically) demonstrated, for making steam and electricity.

From 1945 attention was given to harnessing this energy in a controlled fashion for naval propulsion and for making electricity.

In the course of developing nuclear weapons the Soviet Union and the West had acquired a range of new technologies, and scientists realized that the tremendous heat produced in the process could be tapped either for direct use or for generating electricity. It was also clear that this new form of energy would allow development of compact long-lasting power sources, which could have various applications, not least for shipping, and especially in submarines.

The first nuclear reactor to produce electricity (albeit a trivial amount) was the small Experimental Breeder Reactor (EBR-1) in Idaho, in the USA, which started up in December 1951.

In 1953 President Eisenhower proposed his "Atoms for Peace" programme, which reoriented significant research effort towards electricity generation and set the course for civil nuclear energy development in the USA.

In the Soviet Union, work was under way at various centres to refine existing reactor designs and develop new ones. The existing graphite-moderated channel-type plutonium production reactor was modified for heat and electricity generation, and in 1954 the world's first nuclear-powered electricity generator began operation in the then closed city of Obninsk at the Institute of Physics and Power Engineering (FEI). The AM-1 (Atom Mirny – peaceful atom) reactor was water-cooled and graphite-moderated, with a design capacity of 30 MWt or 5 MWe. It was similar in principle to the plutonium production reactors in the closed military cities and served as a prototype for other graphite channel reactor designs, including the Chernobyl-type RBMK (reaktor bolshoi moshchnosty kanalny – high power channel reactor) reactors. AM-1 produced electricity until 1959 and was used until 2000 as a research facility and for the production of isotopes.

Also in the 1950s Obninsk was developing fast breeder reactors (FBRs). In 1955 the BR-1 (bystry reaktor – fast reactor) fast neutron reactor began operating. It produced no power but led directly to the BR-5, which started up in 1959 with a capacity of 5MWt; BR-5 was used to do the basic research necessary for designing sodium-cooled FBRs. It was upgraded and modernized in 1973 and then underwent major reconstruction in 1983 to become the BR-10 with a capacity of 8 MWt; it is now used to investigate fuel endurance, to study materials and to produce isotopes.

The main US effort was under Admiral Hyman Rickover, which developed the Pressurized Water Reactor (PWR) for naval (particularly submarine) use. The PWR used enriched uranium oxide fuel and was moderated and cooled by ordinary (light) water. The Mark 1 prototype naval reactor started up in March 1953 in Idaho, and the first nuclear-powered submarine, USS *Nautilus*, was launched in 1954. In 1959 both USA and USSR launched their first nuclear-powered surface vessels.

The Mark I reactor led to the US Atomic Energy Commission building the 90 MWe Shippingport demonstration PWR reactor in Pennsylvania, which started up in 1957 and operated until 1982.

Since the USA had a virtual monopoly on uranium enrichment in the West, British nuclear power development took a different tack and resulted in a series of reactors fuelled by natural uranium metal, moderated by graphite, and gas-cooled. The first of these 50 MWe "Magnox" types, Calder Hall-1, started up in 1956 and ran until 2003. However, after 1963 (and 26 units) no more were commenced. Britain next embraced the Advanced Gas-Cooled Reactor (using enriched oxide fuel) before conceding the pragmatic virtues of the PWR design.

9.9 NUCLEAR ENERGY GOES COMMERCIAL

In the USA, Westinghouse designed the first fully commercial PWR – Yankee Rowe (250 MWe), which started up in 1960 and operated to 1992. Meanwhile the boiling water reactor (BWR) was developed by the Argonne National Laboratory, and the first one, Dresden-1 of 250 MWe, designed by General Electric, was started up earlier in 1960. A prototype BWR, Vallecitos, ran from 1957 to 1963. By the end of the 1960s, orders were being placed for PWR and BWR reactor units of more than 1000 MWe.

Canadian reactor development headed down a quite different track, using natural uranium fuel and heavy water as a moderator and coolant. The first unit started up in 1962. This "CANDU" design has been exported, and continues to be refined.

France started out with a gas-graphite design similar to Magnox, and the first reactor started up in 1956. Commercial models operated from 1959. It then settled on three successive

Photograph supplied by UK Nuclear Decommissioning Authority

Calder Hall, UK: One of the world's first nuclear power reactors, which operated for nearly 50 years.

generations of standardized PWRs, which was a very cost-effective strategy.

In 1964 the first two Soviet nuclear power plants were commissioned. A 100 MWe boiling water graphite channel reactor began operating in Beloyarsk (Urals). In Novovoronezh (Volga region) a new design – a small (210 MWe) pressurized water reactor (PWR) known as a VVER (veda-vodyanoi energetichesky reaktor – water cooled power reactor) was built.

The first large (1000 MWe) RBMK (high-power channel reactor) started up at Sosnovy Bor near Leningrad in 1973, and the same year saw the commissioning of the first of four small (12 MW) boiling water channel-type units in the eastern Arctic town of Bilibino for the production of both power and heat.

In the Arctic north-west a slightly bigger VVER with a rated capacity of 440 MWe began operating, and this became a standard design, subsequently enlarged to 1000 MWe.

In Shevchenko in Kazakhstan the world's first commercial prototype FBR (the BN-350) started up in 1972 (BN = bystry neutron – fast neutron), producing 120 MW of electricity and heat to desalinate Caspian seawater. A prototype BOR-60 – had started at Obninsk in 1959, generating 12 MW of electricity.

Around the world, with few exceptions, other countries have chosen light-water designs for their nuclear power programmes, so that today 65% of world capacity is PWR, and 23% BWR.

9.10 THE NUCLEAR POWER BROWN-OUT

From the late 1970s to about 2002 the nuclear power industry suffered some decline and stagnation. Many reactor orders from the 1970s were cancelled. The few new reactors that were ordered, coming on line from mid 1980s, little more than matched retirements. Against this, capacity increased by nearly one third and output increased 60% due to capacity plus improved load factors; therefore the share of nuclear in world electricity from the mid-1980s was fairly constant at 16% to 17%.

The industry's stagnation together with an increase in secondary supplies led to a drop in the uranium price. Oil companies which had entered the uranium field bailed out, and there was a consolidation of uranium producers.

By the late 1990s an expansion in nuclear power in Asian countries, such as Japan and S. Korea, ran counter to the trend in the rest of the world. The first third-generation reactor – Kashiwazaki-Kariwa 6, a 1350 MWe Advanced BWR – was commissioned in Japan in 1996-1997.

Leningrad nuclear power plant at Sosnovy Bor

Photograph supplied by Nuclear Engineering International; photographer – James Varley

Control room at Olkiluoto NPP in Finland

9.11 NUCLEAR RENAISSANCE

In the new century several factors have combined to revive the prospects for nuclear power. First is realization of the scale of projected increased electricity demand worldwide, but particularly in rapidly-developing countries. Second is awareness of the importance of energy security, and third is the need to limit carbon emissions due to concern about global warming. Fourth, fossil fuel prices have increased strongly, thus increasing the economic competitiveness of nuclear power.

These factors coincide with the availability of a new generation of nuclear power reactors, and in 2004 the first of the late third-generation units was ordered for Finland – a 1600 MWe European PWR (EPR). It is now under construction. A similar unit is planned for France as the first reactor of a full fleet replacement in that country. In the USA the 2005 Energy Policy Act provided incentives for establishing new-generation power reactors, and by mid 2006 proposals for over 20 large new power reactors had been announced.

APPENDIX 1
Ionizing radiation and how it is measured

The following are four kinds of nuclear radiation:

Alpha particles: These are particles (atomic nuclei) consisting of two protons and two neutrons and are emitted from naturally-occurring heavy elements such as uranium and radium, as well as from some man-made transuranic elements. They are intensely ionizing but can be readily stopped by a few centimetres of air, a sheet of paper, or the human skin. They are only dangerous to people if they are released inside the body. Alpha-radioactive substances are safe if kept in any sealed container, even a plastic bag.

Beta particles: These are either electrons or positrons (therefore of very low mass) emitted by many radioactive elements. They can be stopped by a few millimetres of wood or aluminium. They can penetrate a little way into human flesh but are generally less dangerous to people than gamma radiation. Exposure produces an effect like sunburn, but which is slower to heal. Beta-radioactive substances are also safe if kept in appropriate sealed containers.

Gamma rays: These are high-energy beams almost identical to X-rays and of shorter wavelength than ultraviolet radiation. Emitted in many radioactive decays, they are very penetrating, and need substantial thicknesses of heavy materials such as lead, steel or concrete to shield them. Gamma rays are the main hazard to people dealing with sealed radioactive materials used, for example, in industrial gauges and radiotherapy machines. Doses can be detected by the small badges worn by workers handling any radioactive materials. Gamma activity in a substance (e.g. rock) can be measured with a scintillometer or Geiger counter.

Neutrons: These are mostly released by nuclear fission, and apart from a little cosmic radiation they are seldom encountered outside the core of a nuclear reactor. Fast neutrons are very penetrating as well as (indirectly) strongly ionizing and hence very destructive to human tissue. They can be slowed down (or "moderated") by wood, plastic, or (more commonly) by graphite or water.

X-rays are also ionizing radiation, virtually identical to gamma rays, but not nuclear in origin. **Cosmic radiation** consists of very energetic particles, mostly protons, which bombard the earth from outer space.

It is important to understand that alpha, beta, gamma and X-radiation does not cause the body or any other material to become radioactive.

Units:

The amount of ionizing radiation absorbed in tissue can be expressed in **grays**: 1 Gy = 1 J/kg. However, since neutrons and alpha particles cause more damage per gray than gamma or beta radiation, another unit, the **sievert (Sv)**, is used in setting radiological protection standards. One gray of beta or gamma radiation has 1 Sv of biological effect, 1 Gy of alpha particles has a 20 Sv effect and 1 Gy of neutrons is equivalent to around 10 Sv (depending on their energy).

Total dose is thus measured in sieverts; millisieverts (mSv), one thousandth of a sievert; or microsieverts (μSv), one millionth of a sievert. The rate of dose is measured in milli- or microsieverts per hour (hr) or year (yr). The average natural dose for humans is around 2 mSv/yr. In industry, the maximum annual dose allowed for radiation workers is 20 mSv/yr; in practice, doses are usually kept well below this level.

These levels contrast with those which are known to be harmful to humans: with gamma radiation a short term dose of 1 Sv causes temporary radiation sickness; 5 Sv would kill about half the people receiving it in a month; a burst of 10 Sv would be fatal to all in a matter of days. The 28 radiation fatalities who died within four months of the Chernobyl disaster appear to have received more than 5 Sv in a few days, while those who suffered acute radiation sickness averaged doses of 3.4 Sv.

The becquerel (Bq) is a unit or measure of actual radioactivity in material (as distinct from the radiation it emits, or the human dose from that), with reference to the number of nuclear disintegrations per second (1 Bq = 1 disintegration/sec). Quantities of radioactive material are commonly estimated by measuring the amount of intrinsic radioactivity in becquerels – 1 Bq of radioactive material is that amount which has an average of one disintegration per second, or an activity of 1 Bq.

Older units of radiation measurement continue in use in some literature:

1 gray = 100 rads
1 sievert = 100 rem
1 becquerel = 27 picocuries or 2.7×10^{-11} curies

One curie was originally the activity of one gram of radium-226, and represents 3.7×10^{10} disintegrations per second (Bq).

Radon and radon progeny

The Working Level Month (WLM) has been used as a measure of dose for exposure to radon and in particular, radon decay products (see Appendix 2). One Working Level is approximately equivalent to 3700 Bq/m³ of Rn-222 in equilibrium with its decay products. Exposure to 0.4 WL was the maximum permissible for workers. Continuous exposure during working hours to 0.4 WL would result in a dose of 5 WLM over a full year, corresponding to about 50 mSv/yr whole body dose for a 40-hour week. In mines, an individual worker's dose is kept below 1 WLM/yr (10 mSv/yr), and typically averages half this.

A background radon level of 40 Bq/m^3 indoors and 6 Bq/m^3 outdoors, assuming an indoor occupancy of 80%, is equivalent to a dose rate of 1 mSv/yr and is the average for most of the world's inhabitants.

Some comparative radiation doses:

2 mSv/year	Typical background radiation dose rate in Australia.
3 mSv/year	Typical background radiation to North American public.
3-5 mSv/year	Typical occupational dose rate (above background) to uranium miners in Canada and Australia.
10 mSv/year	Maximum actual dose rate to Australian uranium miners.
20 mSv/year	Current limit for nuclear industry employees (5 year average).
50 mSv/year	Former long-term limit for nuclear industry employees and uranium miners; current maximum limit in single year.
350 mSv in lifetime	Criterion for relocating people after Chernobyl accident.
1000 mSv	As short term dose, likely to cause temporary radiation sickness.
10,000 mSv	As short term whole-body dose, fatal within a few weeks.

APPENDIX 2

Some radioactive decay series showing half-lives

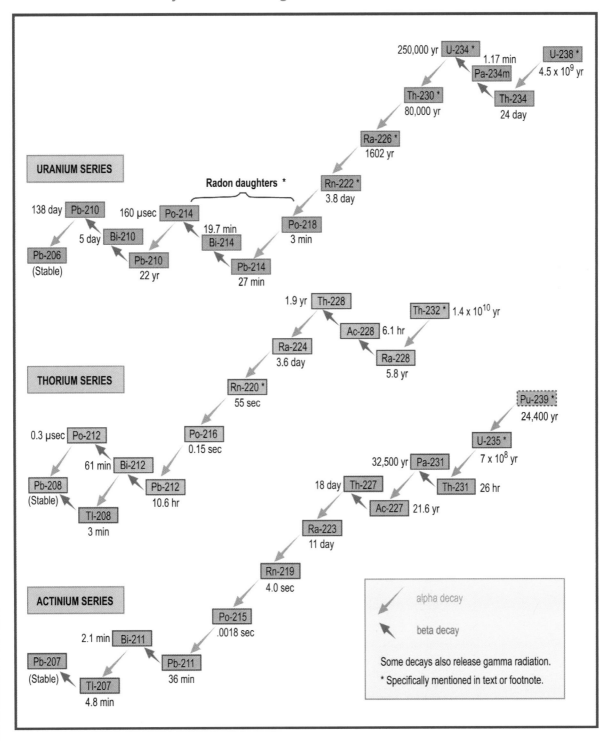

Notes:

1. In a uranium orebody, the U-238 series represents almost 95% of the radioactivity.

2. The level of radiation emitted by an isotope is inversely proportional to its half-life. The shorter the half-life of an isotope, the more radiation it emits per unit mass. Th-232, U-235 and U-238 are thus virtually stable.

APPENDIX 3

Scientific consensus on ethical, societal and technical aspects of radioactive waste management

The following three statements serve to underline the scientific consensus that has been reached on the ethical, societal and technical aspects of radioactive waste management.

The first statement was formulated and published by the International Atomic Energy Agency (IAEA) in 1995 to confront the question of what are the best and most appropriate means of managing and disposing of radioactive wastes from the civil nuclear fuel cycle. The IAEA statement has since become part of the International Waste Convention which entered force in 2001.

The IAEA view was reinforced by a Collective Opinion on Ethical Aspects of Geological Disposal of Long-lived Radioactive Wastes issued by the Organization for Economic Cooperation and Development's (OECD) Nuclear Energy Agency (NEA). The numerous experts from diverse countries involved in preparing this considered that "from an ethical standpoint, including long-term safety considerations, our responsibilities to future generations are better discharged by a strategy of final disposal than by reliance on stores which require surveillance, bequeath long-term responsibilities of care, and may in due course be neglected by future societies whose structural stability should not be presumed." This opinion was endorsed by the IAEA and the European Commission.

A 1999 OECD/NEA statement updates the consensus view on the status of geological disposal and is summarized below.

The third statement gives the principal findings and conclusions of a 2001 US National Academies report on waste management issues.

International Atomic Energy Agency
Fundamental Principles of Radioactive Waste Management

1 – 5 Radioactive waste shall be managed in such a way:

- as to secure an acceptable level of protection for human health.
- as to provide an acceptable level of protection of the environment.
- as to assure that possible effects on human health and the environment beyond national borders will be taken into account.
- that predicted impacts on the health of future generations will not be greater than relevant levels of impact that are acceptable today.
- that will not impose undue burdens on future generations.

156

6. Radioactive waste shall be managed within an appropriate national legal framework including clear allocation of responsibilities and provision for independent regulatory functions.

7. Generation of radioactive waste shall be kept to the minimum practicable.

8. Interdependencies among all steps in radioactive waste generation and management shall be appropriately taken into account.

9. Safety of facilities for radioactive waste management shall be appropriately assured during their lifetime.

IAEA 1995

Geological Disposal of Radioactive Waste
– Review of Developments in the 1990s

In 1999 the Radioactive Waste Management Committee of the OECD NEA surveyed member countries as well as the European Commission and the IAEA to review the adequacy and continuing relevance of earlier collective opinions. A very high level of consensus was found internationally among regulators and implementers. Broad conclusions reached at the end of this review were that:

- Deep geologic disposal concepts have made significant progress in the past ten years, most especially in the technical areas concerning the understanding, characterisation and quantitative modelling of the natural and engineered safety-barrier systems.

- No radical changes in strategy or in applied methodologies have proven to be necessary. Although, refinements are still being made, deep geologic disposal is effectively a technology that is mature enough for deployment.

- In many programmes, more emphasis is being placed upon the contribution of the engineered barriers, but the natural or geologic barriers in a deep repository continue to play a crucial role in determining the achievable long-term safety.

- All national programmes continue to support deep geologic disposal as a necessary and a feasible technology, even though some countries wish to postpone implementation of repositories or to evaluate other options in parallel.

- There is a general common trend towards advocacy of prudent, stepwise approaches at the implementational and regulatory level to allow smaller incremental steps in the societal decision making process. Discrete, easily overviewed steps facilitate the traceability of decisions, allow feedback from the public and/or their representatives, promote the strengthening of public and political confidence in the safety of a facility along with trust in the competence of the regulators and implementers of disposal projects.

- Although one deep geologic repository, purpose-built for long-lived waste, is now operating, the timescales envisioned ten years ago for the development of deep geologic repositories were too optimistic. The delays that have occurred are partly due to operational causes, but mainly reflect institutional reasons, in large part associated with insufficient public confidence.

- There is an acute awareness in the waste management community of this lack of public confidence; efforts are needed by both implementers and regulators to communicate effectively to decision makers and the public their consensus view that safe disposal can be achieved.

- The implementers and regulators are more willing than ever to heed the wishes of the public in so far as these do not compromise the safety of disposal facilities. One common goal is to establish strategies and associated procedures that allow long-term monitoring, with the possibility of reversibility and retrievability. A number of programmes now consider these issues explicitly.

- In spite of the delays, no nation has rescinded its decision to pursue geologic disposal and the consensus for pursuing geologic disposal as the only feasible route for assuring permanent isolation of long-lived wastes from the human environment is unaffected.

Alternative means of radioactive waste disposal have often appeared to have promise prior to consideration of all aspects of the proposal. Several exotic options were studied earlier, and are no longer seriously considered. There are those who, for a variety of reasons, strongly advocate surface storage or partitioning and transmutation. The waste management community does not however, regard extended or indefinite surface storage as a real alternative to geologic disposal; at best it offers a postponement of final disposal. Partitioning and transmutation is also not regarded as an alternative; at best it reduces the volume, or changes the isotope distribution, of wastes requiring disposal.

OECD NEA 1999

Principal Findings and Conclusions from 2001 US National Academies Report on Disposition of High-Level Wastes (HLW)

- **Today's growing inventory of HLW requires attention by national decision makers.** The present situation in the management of radioactive wastes worldwide is one in which—with some important exceptions—safety and security are being achieved by storage, often at or near the facility that produced the waste. Although quantities are minor compared with toxic wastes from other industrial activities, the inventories, particularly of spent fuel, are increasing in many countries beyond the capacity that can be stored in existing facilities. Measures must be taken to deal with this. Moreover, a segment of the public holds concerns and fears that radioactive wastes present an unmanageable threat. The challenge is not just to identify options that are deemed suitable by the technical experts, but also to assure that the decision processes and waste management technologies chosen have broad public support.

- **The only feasible options are storage on or near the earth's surface and geological disposition.** Safe and secure surface storage is technically feasible. The major uncertainty is in the confidence that future societies will continue to monitor and maintain such facilities. It is not prudent to pursue only storage, without development of the geological disposal option.

- **Geological disposal remains the only long-term solution available.** After four decades of study, geological disposal remains the only scientifically and technically credible long-term solution available to meet the need for safety without reliance on active management. It also offers security benefits because it would place fissile materials out of reach of all but the most sophisticated weapons builders. A well-designed repository represents, after closure, a passive system containing a succession of robust safety barriers. Our present civilization designs, builds, and lives with technological facilities of much greater complexity and higher hazard potential.

- **Today the biggest challenges to waste disposition are societal.** Difficulties in achieving public support have been seriously underestimated in the past, and opportunities to increase public involvement and to gain public trust have been missed. Most countries have made major changes in their approach to waste disposition to address the recognized societal challenges. Such changes include initiating decision processes that maintain choice and that are open, transparent, and collaborative with independent scientists, critics, and members of the public.

- **Whether, when, and how to move toward geological disposal are societal decisions for each country.** This decision process will be lengthy, and the time can be used to improve both the technical and the societal bases for these decisions.

- **A stepwise process is appropriate for decision making under technical and social uncertainty.** Some—but not all—of today's uncertainties in predicting the future behavior of a repository system can be reduced or eliminated by further research and development. A stepwise decision process can utilize evolving knowledge to make sound decisions on repository siting (including the geological setting), design, and operation.

- **Successful decision making is open, transparent, and broadly participatory.** National waste disposition programs in democratic countries cannot hope to succeed today without a decision-making process that facilitates choices among competing social goals and ethical considerations. Sufficient time must be devoted to developing this process, including the involvement of broader circles of citizens in examining the choices in an informed way.

- **International cooperation can help achieve national solutions.** Cooperation especially can help less advantaged nations, for example, those with more limited financial means, small nuclear programs, or unfavorable geology. Cooperation can range from shared research programs to shared storage or disposal facilities offered by a host country to other nations. Sharing technology and facilities will reduce the cost burden for all the cooperating nations and will facilitate the establishment of internationally accepted standards. Progress in adopting a solution in one country serves as a positive example to other countries.

APPENDIX 4
Some useful references

World Energy Outlook 2004
OECD International Energy Agency,
Paris 2004
ISBN 92-64-10817-3

Energy for Tomorrow's World – Acting Now!
World Energy Council
London 2000
ISBN 1-901640-06-X

Global Energy Perspectives
Ed. Nakicenovic, Grubler & McDonald for World Energy Council & IIASA
Cambridge University Press 1998
ISBN 0-521-64569-7

Uranium 2003: Resources, Production and Demand. ("Red Book")
Joint report by the OECD Nuclear Energy Agency and the International Atomic Energy Agency 2004
ISBN 92-64-01673-2

Radioactive Waste Management in Perspective
OECD Nuclear Energy Agency
Paris 1996
ISBN 92-64-14692-X

Radiation in Perspective: Applications, Risks and Protection
OECD Nuclear Energy Agency
Paris 1997
ISBN 92-64-15483-3

Chernobyl's Legacy: Health, Environmental & Socio-economic Impacts
The Chernobyl Forum (8 UN agencies and 3 governments)
IAEA Vienna 2005 (51 pp)
no ISBN

Double or Quits? The Global Future of Civil Nuclear Energy
Malcolm Grimston & Peter Beck,
RIIA & Earthscan Publications, London, & Brookings Inst. USA 2002
ISBN 1-85383 913 2 (paper)

Invisible Rays – A History of Radioactivity
G I Brown
Sutton Publishing, Stroud 2002 (248pp)
ISBN 0-7509-2667-8

Radiation and Modern Life
Alan Waltar,
Prometheus Books, New York 2004 (336 pp)
ISBN 1-59102 250 9

All About Nuclear Energy
Bertrand Barre
Areva 2003 (127 pp)
No ISBN

GLOSSARY

The following is a list of terms which are commonly used in the nuclear industry.

Actinide: An element with an atomic number of 89 (actinium) to 102. Usually applied to those elements above 92 (uranium), which are also called transuranics. Actinides are radioactive and typically have long half-lives. They are therefore significant in wastes arising from nuclear fission (e.g. used fuel). They are fissionable in a fast reactor.

Activation product: A radioactive isotope of an element (e.g. in the steel of a reactor core) which has been created by neutron bombardment.

Activity: The number of disintegrations per unit time inside a radioactive source. Expressed in becquerels.

ALARA: As Low As Reasonably Achievable, economic and social factors being taken into account. This is the optimization principle of radiation protection.

Alpha particle: A positively-charged particle from the nucleus of an atom, emitted during radioactive decay. Alpha particles are helium nuclei, with two protons and two neutrons.

Atom: A particle of matter which cannot be broken up by chemical means. Atoms have a nucleus consisting of positively-charged protons and uncharged neutrons of the same mass. The positive charges on the protons are balanced by a number of negatively-charged electrons in motion around the nucleus.

Background radiation: The naturally-occurring ionizing radiation which every person is exposed to, arising from the Earth's crust (including radon) and from cosmic radiation.

Barn: See Cross section.

Base load: The part of electricity demand which is continuous and does not vary over a 24-hour period. Approximately equivalent to the minimum daily load.

Becquerel: The SI unit of intrinsic radioactivity in a material. One Bq measures one disintegration per second and is thus the activity of a quantity of radioactive material which averages one decay per second. (In practice, GBq or TBq are the common units.)

Beta particle: A particle emitted from an atom during radioactive decay. Beta particles may be either electrons (with negative charge) or positrons.

Biological shield: A mass of absorbing material (e.g. thick concrete walls) placed around a reactor or radioactive material to reduce the radiation (especially neutrons and gamma rays) to a level safe for humans.

Boiling water reactor (BWR): A common type of light water reactor (LWR), where water is allowed to boil in the core thus generating steam directly in the reactor vessel (cf. PWR).

Breed: To form fissile nuclei, usually as a result of neutron capture, possibly followed by radioactive decay.

Breeder reactor: See Fast Breeder Reactor and Fast Neutron Reactor.

Burn: Cause to fission.

Burnable poison: A neutron absorber included in the fuel which progressively disappears and compensates for the loss of reactivity as the fuel is used up. Gadolinium is commonly used.

Burnup: Measure of thermal energy released by nuclear fuel relative to its mass, typically gigawatt days per tonne (GWd/tU).

CANDU: Canadian deuterium uranium reactor, moderated and (usually) cooled by heavy water.

Chain reaction: A reaction that stimulates its own repetition, in particular where the neutrons originating from nuclear fission cause an ongoing series of fission reactions.

Cladding: The metal tubes containing oxide fuel pellets in a reactor core.

Concentrate: See Uranium oxide concentrate (U_3O_8).

Control rods: Devices to absorb neutrons so that the chain reaction in a reactor core may be slowed or stopped by inserting them further, or accelerated by withdrawing them.

Conversion: Chemical process turning U_3O_8 into UF_6 in preparation for enrichment.

Coolant: The liquid or gas used to transfer heat from the reactor core to the steam generators or directly to the turbines.

Core: The central part of a nuclear reactor containing the fuel elements and any moderator.

Critical mass: The smallest mass of fissile material that will support a self-sustaining chain reaction under specified conditions.

Criticality: Condition of being able to sustain a nuclear chain reaction.

Cross section: A measure of the probability of an interaction between a particle and a target nucleus, expressed in barns (1 barn = 10^{-24} cm^2).

Decay: Disintegration of atomic nuclei resulting in the emission of alpha or beta particles (usually with gamma radiation). Also the exponential decrease in radioactivity of a material as nuclear disintegrations take place and more stable nuclei are formed.

Decommissioning: Removal of a facility (e.g. a reactor) from service; also the subsequent actions of safe storage, dismantling and making the site available for unrestricted use.

Delayed neutrons: Neutrons released by fission products up to several seconds after fission. These enable control of the fission in a nuclear reactor.

Depleted uranium: Uranium having less than the natural 0.7% U-235. As a by-product of enrichment in the fuel cycle it generally has 0.25-0.30% U-235, the rest being U-238. Can be blended with highly-enriched uranium (e.g. from weapons) to make reactor fuel.

Deuterium: Known as "heavy hydrogen", a stable isotope has one proton and one neutron in the nucleus. It occurs in nature as 1 atom to 6500 atoms of normal hydrogen. (Hydrogen atoms contain one proton and no neutrons.)

Disintegration: Natural change in the nucleus of a radioactive isotope as particles are emitted (usually with gamma rays), making it a different element.

Dose: The energy absorbed by tissue from ionizing radiation measured in grays or sieverts. One gray is one joule per kg, but this is adjusted according to the effect of different kinds of radiation. The sievert is the unit of dose equivalent used in setting exposure standards. (see Appendix 1)

Element: A chemical substance that cannot be divided into simple substances by chemical means; atomic species with same number of protons.

Enriched uranium: Uranium in which the proportion of U-235 (to U-238) has been increased above the natural 0.7%. Reactor-grade uranium is usually enriched to about 3.5% U-235; weapons-grade uranium is more than 90% U-235.

Enrichment: Physical process of increasing the proportion of U-235 to U-238 (see also SWU).

Fast breeder reactor (FBR): A fast neutron reactor (q.v.) configured to produce more fissile material than it consumes, using fertile material such as depleted uranium in a blanket around the core.

Fast neutron: Neutron released during fission, travelling at very high velocity (20,000 km/s) and having high energy (c. 2 MeV).

Fast neutron reactor: A reactor with no moderator and that therefore uses fast neutrons. It normally burns plutonium while producing fissile isotopes in fertile material, such as depleted uranium (or thorium).

Fertile (of an isotope): Capable of becoming fissile by capturing neutrons, possibly followed by radioactive decay. Examples: U-238, Th-232, Pu-240.

Fissile (of an isotope): Capable of capturing a slow (thermal) neutron and undergoing nuclear fission, such as U-235, Pu-239, U-233, Pu-241.

Fissionable (of an isotope): Capable of undergoing fission: if fissile, by slow neutrons; otherwise, by fast neutrons.

Fission: The splitting of a heavy nucleus into two, accompanied by the release of a relatively large amount of energy and usually one or more neutrons. It may be spontaneous but is usually due to a nucleus absorbing a neutron and thus becoming unstable.

Fission products: Daughter nuclei resulting either from the fission of heavy elements such as uranium, or the radioactive decay of those primary daughters. Usually highly radioactive.

Fossil fuel: A fuel based on carbon presumed to be originally from living matter (e.g. coal, oil and gas). Burned with oxygen to yield energy.

Fuel assembly: Structured collection of fuel rods or elements, the unit of fuel in a reactor.

Fuel fabrication: Making reactor fuel assemblies. Sintered UO_2 or MOX pellets are inserted into zircaloy tubes, which are then known as fuel rods or fuel elements. Many of these comprise the fuel assembly.

Gamma rays: High energy electromagnetic radiation from the atomic nucleus, virtually identical to X-rays.

Genetic mutation: Sudden change in the chromosomal DNA of an individual gene, which may produce inherited changes in descendants. Mutation in some organisms can be made more frequent by irradiation (though this has never been demonstrated in humans).

Giga: One billion units (e.g. 1 gigawatt = 10^9 watts or a million kW).

Graphite: Crystalline carbon used in very pure form as a moderator, principally in gas-cooled reactors, but also in Soviet-designed RBMK reactors.

Gray: The SI unit of absorbed radiation dose, one joule per kilogram of tissue.

Greenhouse gases: Radiative gases in the Earth's atmosphere which absorb long-wave heat radiation from the Earth's surface and re-radiate it, thereby warming the Earth. Carbon dioxide and water vapour are the main ones.

Half-life: The period required for half of the atoms of a particular radioactive isotope to decay and become an isotope of another element.

Heavy water: Water containing an elevated concentration of molecules with deuterium ("heavy hydrogen") atoms.

Heavy water reactor (HWR): A reactor which uses heavy water as its moderator (e.g. CANDU).

High-level wastes: Extremely radioactive fission products and transuranic elements (usually other than plutonium) in used nuclear fuel. They may be separated by reprocessing the used fuel, or the spent fuel containing them may be regarded as high-level waste.

Highly (or High)-enriched uranium (HEU): Uranium enriched to at least 20% U-235. (That in weapons is about 90% U-235.)

In situ leaching (ISL): The recovery by chemical leaching of minerals from porous orebodies without physical excavation. Also known as solution mining.

Ion: An atom that is electrically-charged because of loss or gain of electrons.

Ionizing radiation: Radiation (including alpha particles) capable of breaking chemical bonds, thus causing ionization of the matter through which it passes and damage to living tissue.

Irradiate: Subject material to ionizing radiation. Material irradiated by alpha, beta or gamma rays does not become radioactive itself. Irradiated fuel and reactor components have been subject to neutrons in the core and do become radioactive.

Isotope: An atomic form of an element having a particular number of neutrons. Different isotopes of an element have the same number of protons but different numbers of neutrons and hence different atomic mass (e.g. U-235, U-238). Some isotopes are unstable and decay (q.v.) to form isotopes of other elements.

Light water: Ordinary water (H_2O) as distinct from heavy water.

Light water reactor (LWR): A common nuclear reactor cooled and usually moderated by ordinary water.

Low-enriched uranium: Uranium enriched to less than 20% U-235. (That in power reactors is usually 3.5%-5.0% U-235.)

Megawatt (MW): A unit of power equal to = 10^6 watts. MWe refers to electric output from a generator, MWt to thermal output from a reactor or heat source (e.g. the gross heat output of a reactor itself, typically three times the MWe figure).

Metal fuels: Natural uranium metal as used in a gas-cooled reactor.

Micro: One millionth of a unit (e.g. one microsievert is 10^{-6} Sv).

Milling: Process by which minerals are extracted from ore, usually at the mine site.

Mixed oxide fuel (MOX): Reactor fuel consisting of both uranium and plutonium oxides. The Pu usually comprises about 5% and is the main fissile component.

Moderator: A material such as light or heavy water or graphite used in a reactor to slow down fast neutrons by collision with lighter nuclei so as to expedite further fission.

Natural uranium: Uranium with an isotopic composition as found in nature, containing 99.3% U-238, 0.7% U-235 and a trace of U-234. Can be used as fuel in heavy water-moderated reactors.

Neutron: An uncharged elementary particle found in the nucleus of every atom except hydrogen. Solitary mobile neutrons travelling at various speeds originate from fission reactions. Slow (thermal) neutrons can in turn readily cause fission in nuclei of "fissile" isotopes (e.g. U-235, Pu-239, U-233); and fast neutrons can cause fission in nuclei of "fertile" isotopes, such as U-238, Pu-239. Sometimes atomic nuclei capture neutrons without fission.

Nuclear reactor: A device in which a nuclear fission chain reaction occurs under controlled conditions so that the heat yield can be harnessed or the neutron beams utilized. Nearly all commercial reactors are thermal reactors, using a moderator to slow down the neutrons.

Nuclide: Elemental matter made up of atoms with identical nuclei, therefore with the same atomic number and the same mass number (equal to the sum of the number of protons and neutrons).

Oxide fuels: Enriched or natural uranium in the form of the oxide UO_2, used in many types of reactor.

Plutonium: A transuranic element, formed in a nuclear reactor by neutron capture. It has several isotopes, some of which are fissile and some of which undergo spontaneous fission, releasing neutrons. Weapons-grade plutonium is produced in special reactors to give >90% Pu-239; reactor-grade plutonium contains about 30% non-fissile isotopes. About one third of the energy in a light water reactor comes from the fission of Pu-239, and this is the main isotope of value recovered from reprocessing used fuel.

Pressurized water reactor (PWR): The most common type of light water reactor (LWR), it uses water at very high pressure in a primary circuit and steam is formed in a secondary circuit.

Radiation: The emission and propagation of energy by means of electromagnetic waves or particles (cf. ionizing radiation).

Radioactivity: The spontaneous decay of an unstable atomic nucleus, giving rise to the emission of radiation.

Radionuclide: A radioactive isotope of an element.

Radiotoxicity: The adverse health effect of a radionuclide due to its radioactivity.

Radium: A radioactive decay product of uranium often found in uranium ore. It has several radioactive isotopes. Radium-226 decays to radon-222.

Radon (Rn): A heavy radioactive gas given off by rocks containing radium (or thorium). Rn-222 is the main isotope.

Radon daughters: Short-lived decay products of radon-222 (Po-218, Pb-214, Bi-214, Po-214).

Reactor pressure vessel: The main steel vessel containing the reactor fuel, moderator and coolant under pressure.

Repository: A permanent disposal place for radioactive wastes.

Reprocessing: Chemical treatment of used reactor fuel to separate uranium and plutonium and possibly transuranic elements from the small quantity of fission product wastes products and transuranic elements, leaving a much reduced quantity of high-level waste (which today includes the transuranic elements) (cf. Waste, HLW).

Separative Work Unit (SWU): This is a complex unit which is a function of the amount of uranium processed and the degree to which it is enriched (i.e. the extent of increase in the concentration of the U-235 isotope relative to the remainder). The unit is strictly Kilogram Separative Work Unit, and it measures the quantity of separative work (indicative of energy used in enrichment) when feed and product quantities are expressed in kilograms.

For example, to produce one kilogram of uranium enriched to 3.5% U-235, 4.3 SWU is required if the plant is operated at a tails assay of 0.30%, or 4.8 SWU if the tails assay is 0.25% (thereby requiring only 7.0 kg instead of 7.8 kg of natural U feed).

About 100,000-120,000 SWU is required to enrich the annual fuel loading for a typical 1000 MWe light water reactor. Enrichment costs are related to electrical energy used. The gaseous diffusion process consumes some 2400 kWh per SWU, while gas centrifuge plants require only about 60 kWh/SWU.

Sievert (Sv): Unit indicating the biological damage caused by radiation. One joule of beta or gamma radiation absorbed per kilogram of tissue has 1 Sv of biological effect; 1 J/kg of alpha radiation has a 20 Sv effect and 1 J/kg of neutrons has a 10 Sv effect.

Spallation: The abrasion and removal of fragments of a target which is bombarded by protons in an accelerator. The fragments may be protons, neutrons or other light particles.

Spent fuel: Used fuel assemblies removed from a reactor after several years' use; can be reprocessed or treated as waste.

Stable: Incapable of spontaneous radioactive decay.

Tailings: Ground rock remaining after particular ore minerals (e.g. uranium oxides) are extracted.

Tails: Depleted uranium (cf. enriched uranium), with about 0.3% U-235.

Thermal reactor: A reactor in which the fission chain reaction is sustained primarily by slow neutrons, and hence requiring a moderator (cf. Fast neutron reactor).

Transmutation: Changing atoms of one element into those of another by neutron bombardment, causing neutron capture and/or fission. In an ordinary (thermal) reactor, neutron capture is the main event; in a fast neutron reactor, fission is more common and therefore it is best for dealing with actinides. Fission product transmutation is by neutron capture.

Transuranic element: A very heavy element formed artificially by neutron capture and possibly subsequent beta decay(s). Has a higher atomic number than uranium (92). All are radioactive. Neptunium, plutonium, americium and curium are the best known.

Uranium (U): A mildly radioactive element with two isotopes which are fissile (U-235 and U-233) and two which are fertile (U-238 and U-234). Uranium is the basic fuel of nuclear energy.

Uranium hexafluoride (UF_6): A compound of uranium which is a gas above 56°C and is thus a suitable form in which to enrich the uranium.

Uranium oxide concentrate (U_3O_8): The mixture of uranium oxides produced after milling uranium ore from a mine. Sometimes loosely called yellowcake. It is khaki in colour and is usually represented by the empirical formula U_3O_8. Uranium is sold in this form.

Vitrification: The incorporation of high-level wastes into borosilicate glass, to make up about 14% of it by mass. It is designed to immobilize radionuclides in an insoluble matrix ready for disposal.

Waste: High-level waste (HLW) is highly radioactive material arising from nuclear fission. It can be what is left over from reprocessing used fuel, though some countries regard used fuel itself as HLW. It requires very careful handling, storage and disposal.

Low-level waste (LLW) is mildly radioactive material usually disposed of by incineration and burial.

Yellowcake: Ammonium diuranate, the penultimate uranium compound in U_3O_8 production, but the form in which the mine product was sold until about 1970. See also Uranium oxide concentrate.

Zircaloy: Zirconium alloy used as a tube to contain uranium oxide fuel pellets in a reactor fuel assembly.

INDEX